W9-DCU-497

A HISTORY OF DAMS – THE USEFUL PYRAMIDS

Tot aquarum tam multis necessariis molibus pyramidas videlicet otiosas compares...

With such an array of indispensable structures carrying so many waters compare, if you will, the idle pyramids...

S.J. Fontinus (35-104 AD) 'Two books on the Aqueducts of Rome', book 1, paragraph 16.

A History of Dams

The useful pyramids

By
NICHOLAS J.SCHNITTER
Civil Engineer, Zürich, Switzerland

A.A.BALKEMA / ROTTERDAM / BROOKFIELD / 1994

ABB 8875

Published by
A.A.Balkema, P.O.Box 1675, 3000 BR Rotterdam, Netherlands
A.A.Balkema Publishers, Old Post Road, Brookfield, VT 05036, USA

ISBN 90 5410 149 0

Dedication

Dedicated to my father who not only put me on the technical but also on the historical track. Special thanks go to Mr William Abel, Rüschlikon/ZH for his editorial support and to Mr Josef Zeller, Wettingen/AG who traced most of the drawings.

Contents

Foreword

Exalted in their quest for eternity and the presence of the Gods and enigmatic down to the present day, the pyramids of Egypt were considered the greatest works of man for thousands of years. To the critical and profane mind of the Roman Sextus J. Frontinus (35-104 AD), on the other hand, they had the disadvantage of serving no useful purpose at all in comparison with the life-giving master-pieces of Roman hydraulic engineering, which he goes on to describe with all the greater pride. And the same criticism could be levelled against the great defensive structures built by man over the ages, including the awe-inspiring Great Wall of China – wonders of the world indeed, but useless wonders. Wonders that never fulfilled their purpose and often quickly fell into decline, serving posterity only as convenient quarries.

Not so our large dams. Many of them have long surpassed the pyramids of the New and the Old Worlds even in terms of sheer size. The volume of concrete used fifty years ago for the Grand Coulee Dam, for example, would be enough to provide the blocks needed to build three Cheops pyramids, while the embankment volume of the Tarbela Dam is no less than fifty-six times greater, and the Nurek and Grand Dixence Dams are twice as high. And unlike the pyramids, they were built to offer multiple benefits throughout the world. Benefits which have become literally vital to the human race as it heads towards a population total of six billion. The title of this book is therefore apt indeed.

Dams are built to impound water and discharge it for a variety of purposes into channels, tunnels and conduits, to create the head of water needed for hydropower generation, and to protect estuaries against the ravages of flood tides. Above all, however, they provide active storage for the management of water, that is for striking a balance between natural flow regimes on the one hand and patterns of demand which are normally quite different, if not diametrically opposed, on the other. Depending on the size of the reservoir, this equalizing function may occupy a daily, weekly or annual cycle, or even tide over periods like the scriptural seven lean years.

The *World Register of Dams*, published by the International Commission on Large Dams (ICOLD) currently lists over 36 000 'large dams' around the world, and this figure – impressive as it already is – would greatly increase with every metre by which the arbitrary height limit of 15 m for inclusion in the register is reduced. With only very few exceptions, all these dams meet in full the expectations placed in them. They offer security against two extremes: Against a lack of water bringing drought, power failures, dried out river beds and falling groundwater levels, and against too much water, especially too much too quickly, in the form of raging floods causing devastating inundation to farmland and people's homes. Thus, the multipurpose benefits of storage reservoirs have attained an impressive order of magnitude in every respect.

Their primary function worldwide is to provide reliable water supplies to a major proportion of the approximately 2 700 000 km^2 presently under irrigation in the world. And for the future the United Nations Development Programme (UNDP) requires a 3% compound rate of annual growth to meet the needs of an extra one billion people in the next ten years.

The control of floods, which still cause 40% of all fatalities from natural catastrophes worldwide, has always been a particularly significant motive for dam construction everywhere, and sometimes its primary purpose. In many cases successful permanent flood protection has been achieved for vast areas of land with huge multipurpose projects – from the famous Hoover Dam and the dams built by the Tennessee Valley Authority to the more recent dams built or planned to control the great and dangerous rivers of China. In many other cases flood control is a valuable spin-off from dams built primarily to serve other purposes which also represent the subsequent source of revenue, in particular power production.

With a total annual generation of 2.1 million GWh, hydropower today account for 20% of electricity consumption and about 7% of total energy consumption worldwide. Even at a conservative estimate, the total exploitable potential is at least six times larger – an incentive for the further development of this source of clean, renewable energy, and a challenge to dam engineering.

Storage reservoirs form the backbone of innumerable water supply systems, and many more of them will still be needed for this purpose. Efforts to supply water have not kept up with the unprecedented growth in population, in particular in the developing countries, where so many already face a serious scarcity of water or are still without safe water.

Augmentation of low flows in rivers for navigation purposes, water withdrawals or the improvement of water quality, prevention of river-bed degradation, the enhancement of tourism, recreation areas, fishery, and even bird and wildlife sanctuaries – they are all additional assets rightly claimed for so many dams and their reservoirs.

At present reservoirs behind dams store some 5500 km^3 of water, of which about 3600 km^3 is available for regulated use, thus adding about 25% to the stable portion of annual river runoff in the entire world. This figure is of the same magnitude as the world-wide water withdrawals for consumptive use, estimated at present at about 3700 km^3/year, of which 68% is used for the irrigation of croplands, 24% for industry, and 8% for domestic and industrial purposes.

The sheer magnitude of these few figures shows how much dams and appurtenant reservoirs have become an integral part of our large-scale engineered infrastructure, of our man-made basis of survival. At the same time there can be no doubt that today's awesome rate of growth in the world population, and the pressures that growth is creating in terms of water, food and energy supplies, and need of flood protection, require the continued creation and maintenance of further storage volumes. This need, to be met with due regard for environmental and social compatibility, will continue to challenge the world's dam engineers.

In meeting this challenge we have the advantages of a mature technology, and a vast store of knowledge and experience at our disposal. Progress in the art and science of dam engineering has been enormous, and as a product of the common endeavour of so many dam engineers all over the world this progress still continues. This is clearly reflected in the vast literature on the present state-of-the-art and the trend of future developments, which centres on the work of the International Commission on Large Dams and is already almost too much for the individual to keep abreast of.

This book, however, approaches the subject from quite a different angle by investigating the past. It tells the story of the evolution of dam engineering, which has justly been called, 'one of the most essential aspects of man's attempt to harness, control and improve his environment' (Smith 1971).

It is true that about 85% of all dams were built after 1950 (since when the world population has more than doubled), and most of the older dams date back over the past 180 years or so that have elapsed since the beginning of the Industrial Revolution. But the actual roots of dam engineering can be traced back over a further five thousand years to the beginnings of early civilizations, which often were initiated and enhanced by such community tasks as the construction of dams and canals.

This long era of early history in dam engineering finishing with the Industrial Revolution, in which progress still had to be made on a purely empirical basis and without the accelerating impact of science and technology, is the subject of about two-thirds of the book, in five clearly structured chapter devoted to the various periods and regions of the world. They contain a wealth of highly interesting information as well as many surprising facts like the following:

The first dam for which there is reliable evidence was built in Jordan five thousand years ago to supply the city of Jawa with drinking water, but there may have been even earlier works in Armenia. Around the year 1800 BC, at the time of the pharao Amenemhet III, the Egyptians constructed a reservoir known as the Lake of Moeris, which was in operation for no less than 3600 years with a storage capacity later upgraded to an astounding 275 million m^3. Similarly Yemen is the site of the 20 m high Marib dam, which was built 2500 years ago and is known to have been used for irrigation purposes for over 1300 years before it was abandoned, still to be seen today as a ruin complete with intake works and spillways, downstream from a much larger modern dam that was built just a few years ago. And to the Romans, those masterbuilders of antiquity, we owe the precursors of out modern buttress, arch and multiple arch dams; the Proserpina and Cornalvo embankment dams built by them in Spain are still in use today.

In the Middle Ages, too, some of the achievements in dam engineering were astounding, and have only been surpassed in this our twentieth century, such as the 60 m high Kurit arch dam built by the Mongolians in Iran around 1350, the 13.6 km long Pollonaruwa dam, with an embankment volume of 4.6 million m^3, built under the Singhalese kings in Sri Lanka around 1170, or the Hongze reservoir in China, with a storage volume of 13 km^3, which was built some time after 1570. A number of slightly more recent – but equally impressive – dams were built in Europe, like the 46 m high arched gravity dam of Tibi in Spain in 1594, or the 36 m high St. Ferréol dam, an earth dam with a masonry core built in France in 1675.

The sixth chapter is devoted to the evolution of modern dam technology, and this comprehensive subject not surprisingly occupies the remaining one third of the book. It deals with a period covering only two centuries, but a period in which the unprecedented rise of science and technology combined with likewise revolutionary developments in powerful construction machinery to trigger an incredible surge of activity in dam engineering in an economic climate of hitherto unheard-of dynamism and capital investment. It traces the main lines of development of the various types of dams, of the art of foundation treatment and of the design of spillways and outlets, relating them to the names of great engineers and landmark projects. The result is a clearly arranged chronicle of the pioneering ideas and step by step progress made in basic theories and mathematical analyses, in investigation methods and construction methods, and measuring instruments and monitoring systems, right up to the present day.

A final commentary on the present state-of-the-art and a short outlook on the future of dam engineering round off a book that closes a lacuna in the literature that has long been felt. True, there have been many historical works in the field devoted to certain epochs, civilizations, countries, regions or even continents, which doubtlessly served the author well as an indispensable basis and as highly valuable sources of information, but to my knowledge there has never before been a global account of the full history of dam engineering.

The author is clearly well qualified to fill the gap. Born the son of professor Dr.h.c. Gerold Schnitter (1900-1987), who was equally successful and internationally esteemed as a building contractor, expert engineer and university professor, and was also Vice-President of the International Commission on Large Dams, Nicholas J. Schnitter (born 1927) literally grew up in the world of dam engineering. Having studied civil engineering himself at the Swiss Federal Institute of Technology in Zürich, he has been active in the field for 36 years, including long years of service as a senior executive at Motor Columbus Consulting Engineers Ltd., Baden Switzerland, and has played a key role in a number of major projects, including such very high arch dams as Zervreila and Emosson in Switzerland or El Cajon in Honduras. In addition to the expert knowledge of a true professional, N.J. Schnitter has always shown a keen interest in history, and has taken advantage of his many travels to further and stimulate a hobby that has equipped him so well to finally write his book.

Five thousand years of dams is such an important thread in the colourful fabric of engineering history that the book will be sure to delight a wide readership well beyond the narrower limits of the profession. And the dam engineers themselves are particularly grateful to N.J. Schnitter for having compiled and written this fine history of their art and science from among their ranks.

Dr.Dr.h.c. Wolfgang Pircher
President, International Commission on Large Dams

Conversion factors

International Standard Unit (one)		American Units		
Abbrev.	Designation	Factor	Abbrev.	Designation
kg	Kilogram	2.21	lbs	Pounds
kg/m^3	Kilogram per cubic metre	0.06	lb/cf	Pound per cubic foot
	(for cement content:	0.02	bag/cy	Bag per cubic yard)
km	Kilometre	0.62	mi.	Mile
km^2	Square kilometre	0.39	sq. mi.	Square mile
km^3	Cubic kilometre	0.81	10^6ac-ft	Million acre-feet
m	Metre	3.28	ft	Feet
m^2	Square metre	10.8	sq. ft	Square feet
m^3	Cubic metre	1.31	cy	Cubic yard
m^3/d	Cubic metre per day	264	gpd	Gallons per day
m^3/s	Cubic metre per second	35.3	cfs	Cubic feet per second
Million m^3	Million cubic metres	811	ac-ft	Acre-feet
MN	Meganewton	112	t	Tons (force)
MPa	Megapascal	145	psi	Pounds per square inch

Inclination in %: horizontal/vertical.
Slope x: y: vertical: horizontal.

Figure 1. The only dam-building animal besides man is the beaver, and he has been at it much longer! Beaver dam on the Averill creek near Merrill in Wisconsin/USA (Photo by the author).

CHAPTER 1

The ancient civilizations

1.1 THE NILE VALLEY

The 'father of historiography', Herodotos of Halikarnassos (484–
425 BC), already stated that Egypt is a gift of the river Nile
(Herodotos). In fact, only its yearly flood from August to November rendered agriculture possible in the desert area (average yearly
precipitation in Cairo 20 mm). From about 5000 until 3000 BC
agriculture was practised without specific irrigation systems. The
earliest evidence for such structures dates from around 3100 BC
(Fig. 2) (Butzer 1976). These did not include any major diversion
weirs or dams, although Herodotos mentioned, that the founder
of the first dynasty of kings, Narmer (in Greek: Menes, about
2900 BC) built one at Kosheish near the capital city of Memphis,
20 km to the south of Cairo (Herodotos). It appears, however, to
have rather been a training dike to divert the Nile towards the east
for the purpose of protecting the city against flooding.

First hydraulic works

Kosheish dike

1.1.1 *Sadd-el-Kafara*

The ruins of a real water retaining dam (in Arabic: 'sadd' = closure)
in the modern sense were discovered over one hundred years ago
in the Garawi ravine facing Memphis (Fig. 3). They were investigated thoroughly in 1982 (Garbrecht 1985). The dam was built
around 2600 BC, i.e. at the beginning of the Pyramid Age. Accordingly, its dimensions of 14 m in height and 113 m in crest
length were considerable. It is the oldest dam of such size known
so far in the whole world. Purpose of the reservoir of half a million
m^3 capacity was the retention of the rare but violent floods in the
ravine and their subsequent evaporation.

First large dam in the
world

The grossly overdesigned cross section of the dam was probably
due to lacking experience. This is particularly true for the central
impervious core of silty sand and gravel between the two shells of
rockfill (Fig. 4). The total volume of fill thus reached 87 000 m^3

Ample dimensions

Figure 2. The Scorpion King inaugurating an irrigation network in Egypt around 3100 BC (from Butzer 1976).

whereas the construction period took some eight to ten years. The fabrication and the careful placement of the 17 000 revetment blocks on both outer faces of the dam each weighing 300 kg, must have been particularly time consuming (Fig. 5).

Since there was no channel or tunnel to divert the river across or around the site during the long construction period the dam was most probably breached well before completion during one of the flash floods, it was intended to control. The consequences of the dam's failure must have been so grave that Egyptian engineers refrained from further dam constructions for some eight centuries.

Figure 3. View of Kafara dam site from the left bank; US = upstream, DS = downstream (Photo H. Fahlbusch, Ratzeburg/D).

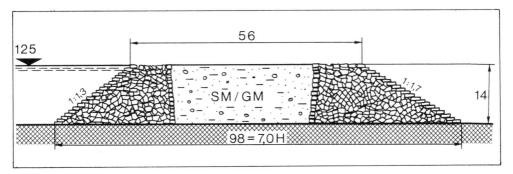

Figure 4. Cross section of Kafara dam (after Garbrecht 1985).

Figure 5. Upstream face of Kafara dam (photo G. Garbrecht, Lagesbüttel/D).

1.1.2 *Semna and Mala'a dams*

King Amenemhet III (in Greek: Moeris, 1842-1798/95 BC) is reported to have had the Nile dammed near Semna (Sudan), 950 km south of Cairo, in order to facilitate navigation across its second cataract, but little is known about this structure (Vercoutter 1980). By contrast, many studies have been devoted to Moeris' artificial lake in the Faiyûm depression, 90 km southwest of Cairo, which Herodotos visited around 445 BC (Herodotos). A thorough review of his and all subsequent reports as well as an extended field survey were undertaken in 1988, the conclusions of which form the basis for the present account (Garbrecht 1990).

Moeris' lake was part of an irrigation scheme using water from the 340 km long Yûsef branch of the Nile by restoring its connection with the Faiyûm depression, which had been interrupted about 7500 BC (Fig. 6). The reservoir was located in the southeastern corner of the depression, where remains of a dam can still be seen. However, these remains do not date from Moeris' times. They rather belong to Ptolemaic (third century BC) and later reconstructions. This dam was over 8000 m long and up to 7 m high. It consisted of a vertical water retaining wall of masonry supported on the downstream side by buttresses and/or an embankment, as later frequently used by Roman engineers (see below) (Fig. 7).

As rebuilt in the third century BC the Mala'a reservoir had an unprecedented capacity of about 275 million m³. If excessive amounts of water flowed into it during the Nile flood they could be spilled into the Gharaq depression to the west. The stored water

Moeris' 19th century BC dams

Huge reservoir in the Faiyûm

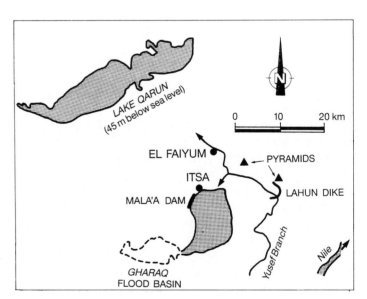

Figure 6. Location map of Mala'a dam and Lâhûn dike (after Garbrecht 1990).

Figure 7. View along the water retaining wall of the Ptolemaic/Roman Mala'a dam; upstream is to the left (photo G. Garbrecht, Lagesbüttel/D).

was released through two multiple outlets early in the following year for a second crop, and, conceivably the emptied reservoir area was cultivated too. Despite several breachings of the dam in its highest part, the Mala'a reservoir was in operation until the end of the 18th century AD, i.e. for over 2000, or, if counted from Moeris' times, for some 3600 years. This certainly represents a world record!

3rd century BC
reconstruction

Although the inflow into the Mala'a reservoir was controlled at its diversion from the Yûsef branch about 5 km southeast of El Faiyûm, an additional regulating structure was installed before the first century BC near Lâhûn some 11 km further to the east in order to control the total inflow to the Faiyûm depression (Fig. 6). From the beginning it must have formed part of a dike against the Nile floods, although it was first mentioned as late as in the 11th century AD. The remains of this dike, which does not follow the shortest possible alignment, but instead bulges towards the Nile valley, today measure some 5000 m in length and up to 4 m height (Fig. 8).

Lâhûn regulating dike

Figure 8. Side facing Nile valley
of the Lâhûn dike; pyramid in
background (photo by the
author).

1.1.3 *Meroitic reservoirs*

Rainfall collection basins
in Kush

From 920 BC to about 350 AD northeastern Sudan was dominated by the kingdom of Kush (Egyptian name), whose capital
was since 591 BC Meroë on the Nile, some 200 km northeast of
Khartoum. For the settlement of the semiarid area (average yearly
precipitation about 150 mm) south of the capital hundreds of so
called 'hafirs' were built (Kleinschroth 1987). They consisted of
embankments in the form of a horseshoe or an open circle, the
opening being located at the foot of a hill to collect its runoff. The
embankment material was excavated from the floor of the reservoir, thus providing additional storage capacity. However, this also
implied a water-lifting device (pole-and-bucket lever known since
1340 BC or a water-screw, bucket-chain and waterwheel since
250 BC) to exhaust the water stored below ground level. Those
amounts of water stored above ground level were let out through
a temporary breach in the embankment which was repaired after
the reservoir had been emptied.

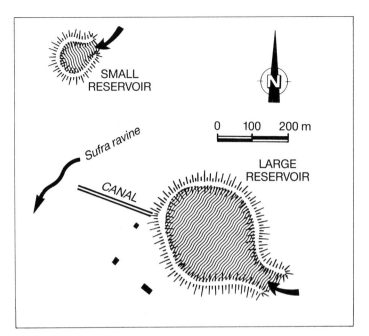

Figure 9. Plan of the Meroitic reservoirs at Musawwarat (after Kleinschroth 1987).

Many of these 'hafirs' were of considerable size and several attained a diameter of 250 m at the crest of the embankment. The largest and one of the oldest (300–250 BC) was located at Musawwarat es Sufra, 130 km northeast of Khartoum (Fig. 9). Its embankment was 800 m long, up to 15 m high. It had a 1:2 slope on both faces (vertical:horizontal). The excavation was also 15 m deep and sloped 1:6 towards the centre. The volume of the material moved must have come close to 250 000 m³ of fill, while the storage capacity exceeded 500 000 m³. The dam crest was depressed in several places to concentrate the spilling of excess inflow, which could not be controlled. Although all embankment slopes were protected by a layer of stones, considerable erosions occurred soon. Nevertheless, the reservoir was in operation for some 600 years under Meroitic rule and then again for another 900 years from about 600 to 1500 AD. In the second period a 60 m long outlet tunnel 0.5 m wide and 1.0 m high was built through the base of the embankment. Its walls consisted of dry masonry and its cover of large stone slabs.

Musawwarat reservoir's...

...structural details

1.2 GREECE

The earliest Greek civilization, that of the Mycenaean heroes of later times, already excelled in hydraulic engineering (Schnitter 1984). Daedalus, the Athenian builder of the famous labyrinth on

Crete (around 1500 BC), is reported to have built a dam in Sicily/Italy after his flight from the island, which cost Icarus, his pert son's life.

1.2.1 *Kofini diversion dam*

River diversion in the
time of Herakles

Another of the ancient heroes, Herakles (around 1300 BC) was the son of the banned king of Tiryns in the Peloponnesus (Fig. 10). For the new ruler, king of nearby Mycenae, he had to perform his famous twelve labours. The fifth concerned the cleaning of the stables of Augeas, which task he accomplished by diverting a river through it. An actual river diversion was undertaken 4 km east of Tiryns to protect the city against the floods of the Lakissa (Balcer 1974). To this end a 3 km long canal was dug in a southeasterly direction and the river forced into it by damming its old bed. Below the end of the canal the river formed a new bed, passing south of the city at a safe distance.

Figure 10. Map of Greece with locations of the ancient dam sites.

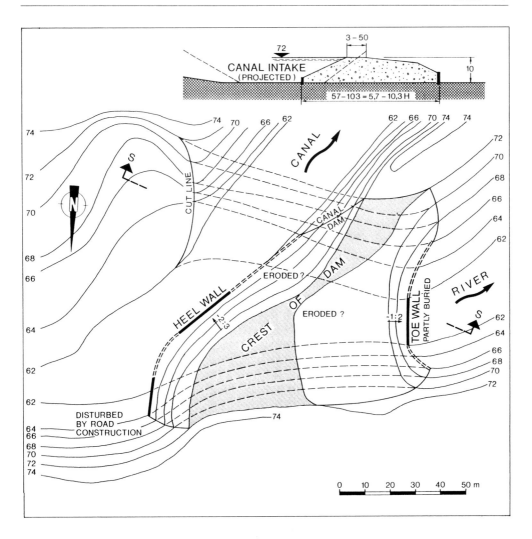

The present dam, which still fulfils its purpose, was built around 1260 BC (Fig. 11). It consists of an embankment between two low walls of cyclopean masonry, of which the upstream one also served to protect the dam's heel against erosion by the river current. The distance between the walls diminished from 103 on the right bank to 57 m on the left side of the valley. The base width of the embankment measuring up to 10 m in height was thus overdesigned on its right side, while it conformed with modern standards on the left bank. The same is true for the width of the 100 m long crest, which was partly eroded from rainwater and slight overtoppings. Heavier incidents of the latter sort did, however, not occur. This means that the discharge capacity of the diversion canal was correctly chosen some 3300 years ago!

Figure 11. Plan and section of the Kofini diversion dam on the Peloponnesus (topography from Jantzen 1975).

Details of diversion dam

1.2.2 *Mycenaean flood control and irrigation dams*

Ingenious utilization of
water logged valleys

In the centre of the Peloponnesus and northwest of Athens there are several valleys, which are drained only through natural galleries within their calcareous underground and therefore they became water logged after the start of the rainy season in October until after the beginning of the following year's agricultural cycle. As a remedy the Mycenaeans halved several of these valleys by dams. The one section containing all or most of the subterranean outlets, which were sometimes enlarged artificially, was cultivated and provided with drainage ditches for rapid dewatering. In the other part of the valley the runoff from rain and snowmelt was stored and used for irrigation during the dry season. A recent review of all reports by travellers since Pausanias (second century AD) and extensive field surveys led to the identification of the eight Mycenaean reservoirs listed in Table 1 (Knauss 1990).

All the dams impounding these reservoirs were of low, but mostly very long and wide design (Fig. 12). Half of them consisted of two masonry walls with an earthfill in between. The walls were

Table 1. Mycenaean reservoirs.

Name (location Fig. 10)	Dam height (m)	Dam length (m)	Reservoir capacity (million m^3)	Dam structure
Boedria	2	1250	24	Earth between walls
Kinéta	2.5	1200	4	Earthdam (toe wall later)
Mantinea	3	300	15	Single wall
Orchomenos	2	2100	16	Earthdam
Permessos	4[1]	200	2	Earth between walls
Pheneos	2.5	2500	19	Wall with gravel support
Stymphalos	2.5	1900	9	Earth between walls
Takka	2	900	9	Earth between walls

[1]Heightened: 1 m in the 10th to 13th century AD.

Figure 12. View along the Mycenaean dam of Stymphalos with remains of a Roman aqueduct laid on its crest about 125 AD; upstream is to the left (photo J. Knauss, Kochel a.S./D).

almost equally wide as they were high. Their outer faces were lined with cyclopean blocks, whereas for the interior the Mycenaeans used rubble masonry with clay as mortar. The walls would thus remain impervious even if they settled into the firm lake deposits, on which they were usually founded. Moreover a layer of clay and gravel was placed between the foundation and the walls. The earthfill between the walls was about two to three times as thick as the dam height.

Two of the eight dams were provided with only one wall which, at Mantinea, was unsupported. The Pheneos dam was backed up by an ample gravel fill on the downstream side. The last two dams were pure embankments, whereof the one at Kinéta was about 5 m wide at its crest and both faces sloped at about 1:3, so that the base width amounted to some 20 m (Knauss 1989). A toe wall of ashlar masonry was added on the downstream side in the course of repairs, that were probably ordered by Alexander the Great (356-323 BC). Also the companion structure on the Permessos in the same valley required reparations after being breached by over-topping. It may well have been restored in Alexander's time as well. Still clearly visible are the repairs and the heightening by 1 m between the 10th and 13th centuries AD. They were carried out with rubble masonry, that proved to be too weak in its breached section, so that it failed again by overtopping (Knauss 1989). The other Mycenaean reservoirs operated for many centuries until they were silted up. Only the Mantinea dam found a premature end by its breaching in 418 BC during the Peloponnesian war, the earliest known bellicose act against a dam (Knauss 1990).

Structure of Mycenaean dams

Various dam types

Bellicose act against dam

1.2.3 *Mycenaean dikes and later dams*

In the Boedria and Orchomenos reservoirs the Mycenaeans uti-lized the shallow ends for agriculture by cutting them off with dikes. Ample use of this technique was made along the shores of the large, natural Lake Kopais, some 70 km northwest of Athens (Fig. 13) (Knauss 1987). The separation of several especially large bays to the northeast and -west of the lake, inevitably reduced its surface and flood storage capacity to a considerable degree. Moreover, the lake was cut off from its main subterranean outflows in the north-east. A canal was consequently built along the northern lake shore to carry the waters of the main tributary, the river Kephissos, directly to the principal outflow, in front of which a temporary retention basin was also arranged. The canal was 25 km long, 40 m wide and 2 to 3 m deep. It was also used for navigation.

The ingenuous system came into operation about 1300 BC and the respective area became prosperous. As about half of the inflow was diverted during the rainy season and all of it during the dry

Reclamation dikes around lake Kopais

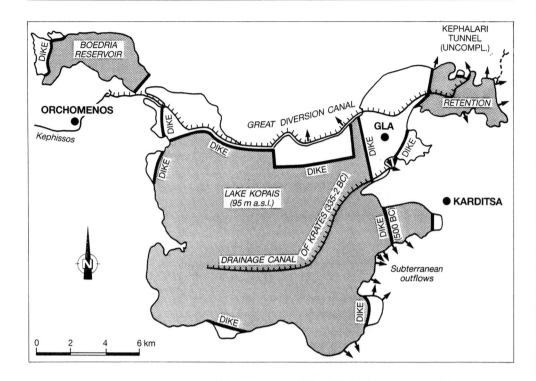

Figure 13. Map of Lake Kopais with the various hydraulic engineering works carried out in and around it in Antiquity (after Knauss 1987).

Figure 14. Cyclopean masonry on the upstream face of the dike southwest of Karditsa from the 5th cent. BC (photo by the author).

Disruption of the Kopais system and later works

season Lake Kopais fell partly dry in summer and could be cultivated. After internal strife, the Trojan War (1218-1209 BC) and following an earthquake the system was disrupted and Lake Kopais reverted back to its original state. In the fifth century BC a new dike was built across the bay southwest of Karditsa (Fig. 13) in the old Mycenaean fashion (Fig. 14). By order of Alexander the Great his engineer Krates of Chalkis attempted from 335 to 332 BC to drain the lake by means of a canal to one of the underground outflows (Fig. 13). It is unknown who started to build the incomplete Kephalari tunnel in the northeast for the same purpose (Fig. 13). Final drainage of the Kopais lake was achieved only in the first third of the present century.

After the Mycenaena exploits almost no dams appear to have been built in ancient Greece. There is a vague reference to two such structures of moderate size, which were built in the fifth century BC near Demoliaki and Kamariza, some 40 km southeast of Athens (Fig. 10) (Kalcyk 1982). They consisted of ashlar masonry and diverted water to the Athenian silver mines in that area. Similarly there was an 11 m high and 25 m long, rapidly silted-up water supply dam dating to the fourth century BC near Kandila on the Greek westcoast (Fig. 10) (Murray 1984). An odd event was reported by Herodotos in connection with the unsuccessful campaign of the Iranian king Xerxes I (486-465 BC) against the Greeks in 480-479 BC (Herodotos). He threatened to drown Thessaly by damming the Pinios river in a gorge near its mouth (Fig. 10). Such destructive dam building was actually carried out in 385-384 BC by the Spartans, who flooded the clay walls of the city of Mantinea in order to destroy them (Murray 1984).

Various small dams in Greece

Destructive dam building

1.3 TURKEY

1.3.1 *Some Hittite remains*

In the 17th century BC the Hittites established the first Indo-European state in the centre of Turkey, with Hattusa (today Boğazkale) as their capital, 150 km east of Ankara. A primitive reservoir embankment was found some 30 km northeast of it at Gölpunar, which might be of Hittite origin (Bittel & Naumann 1952). A small reservoir near Karakuyu, 370 km east of Ankara, dates from the late period of their state in the 13th century BC (Osten et al. 1933). Like the abovementioned, considerably younger Meroitic 'hafirs', it was built at the foot of a hill and partly excavated into it to obtain the material for the embankment (Fig. 15). The latter was some 440 m long and partly lined on the upstream side with flat stones. The remains of a 1.4 m wide sluice were found in the breached section with walls and an upstream closure of flat blocks, two of which show hieroglyphic inscriptions. The water discharged through the sluice was used for the supply of nearby grazing-lands. At the foot of the sacred spring of Eflatun Pinar, 260 km southwest of Ankara, a small pond was impounded by a low dam across the valley (Naumann 1955-1971). The short dam ended on one side in a large block with three outlet notches (Fig. 16).

Hittite rainfall collection basin

Figure 15. Plan of the Hittite reservoir near Karakuyu in Turkey (after Osten et al. 1933).

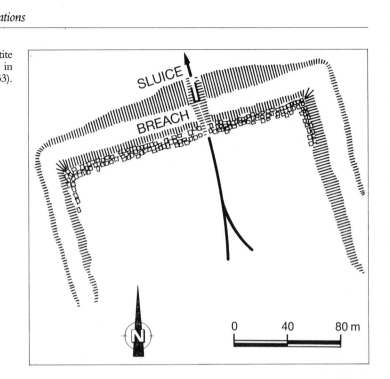

Figure 16. Spillway block of Eflatun Pinar pond in Turkey (photo A. Vogel, Vienna).

1.3.2 *Urartian water supply systems*

Water supply for Tuşpa (Van)

In the 13th century BC the survivors of the kingdom of Mittani (Iraq/Syria/Turkey), destroyed by the Hittites, moved northeast into the area of the large Van lake in southeastern Turkey. Eventually they amalgamated into the new kingdom of Urartu, with its capital city Tuşpa (today : Van) on the southeastern shore of Van Lake (Fig. 17) (Garbrecht 1980). Since the latter's water is undrinkable and the Engusner and Doni rivers flowing through the city do

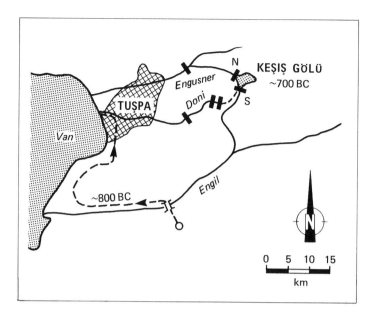

Figure 17. Ancient dams and canals near Tuşpa (Van) in Turkey (after Garbrecht 1980).

not carry water all year round, it was necessary to exploit resources farther away. In a first phase a 56 km long canal was built under king Menua (805-785 BC) from a powerful and steady spring in the Engil valley south of Tuşpa. Thus some 200 000 m³ per day of good water were brought to the city and the surrounding fields. With some modernization and improvements the canal is still in operation after 2800 years!

For the purpose of extending the water supply which became necessary about one century after the construction of Menua's canal a different and in certain respects novel scheme was adopted. It consisted in storing the snowmelt runoff of the Engusner in its wide valley of origin 25 km east and 900 m above the city named Keşiş Gölü (Fig. 17). Later the water was released down the river bed as needed. In order to obtain an adequate storage capacity it was necessary to impound some 100 million m³, which in turn required the damming of not only the Engusner but also of a secondary valley to the south.

Huge mountain storage

The main dam in the north (N in Fig. 17) was destroyed by a flood in 1891 (sic!), so that its type of construction and exact dimensions remain unknown. It was probably some 15 m high and 75 m long with a low wing dam of 200 to 300 m length on the right bank (Garbrecht 1987). The present structure built in 1952 some 50 m upstream of the ancient one is only 5 m high. It has a controlled outlet, which the original dam is believed to also have been equipped with.

Main dam

Clearly interpretable ruins are left of the southern dam (S in Fig. 17 and Fig. 18). It was up to 7 m high and some 60 m long. It

Figure 18. Ruins of the southern dam of Keşiş Gölü reservoir in Turkey; upstream is to the right (photo G. Garbrecht, Lagesbüttel/D).

Secondary dam

consisted of two dry masonry walls with base widths of 7 m each and of an intermediate, up to 13 m wide earth core. The width-to-height rations of the individual sections as well as of the total corresponded thus to one of the Mycenaean dam types described earlier, although on a much larger scale. The same would have been true for the Sadd-el-Kafara, had its outer faces been vertical. The right side of the southern dam had an outlet measuring 0.27 m in breath and 0.90 m in height, which could be closed on the upstream side. Its discharge was deviated to the Doni river by means of a canal for diversion to Tuşpa (Fig. 17).

Other Urartian dams

Near the southern dam a stela was found, on which king Rusa describes and praises in detail the work accomplished during his reign. Unfortunately we do not know, whether he was Rusa I (730-741 BC) or Rusa II (685-645 BC). At a later date several additional dams were built on the Engusner and Doni rivers downstream from Keşiş Gölü (Fig. 17). For lack of archaeological evidence and/or excavations it is at present impossible to properly date them. Remains of other Urartian dams were found near Muradiye, 70 km northeast of Van, and near Adilcevaz, 60 km northwest of it (Garbrecht 1987). While the first one measuring 5 m in height was structured like the southern dam at Keşiş Gölü the second one even had three masonry walls and two earth cores. Its total width amounted to 17 m and its length to 57 m. The height is unknown.

1.4 IRAQ

1.4.1 *The Marduk or Nimrud dam legend*

Like in Egypt, great civilizations developed very early in Iraq on the basis of irrigation. But unlike the high-water in the Nile valley, the yearly floods of Iraq's two main rivers, Euphrates and Tigris, occurred in the wrong season for agriculture and they were often violent. Therefore a combination had to be practised of flood protection measures and water withdrawal at lower river stages by means of irrigation canals. To lead the waters from the rivers into the canals it was neither necessary nor possible with the available technology of damming them up completely. The Tigris in particular was avoided as a source of irrigation water and therefore the Marduk or Nimrud dam allegedly built in 2500 BC near Samarra, 100 km northwest of Baghdad, is really a legend. This is substantiated by the fact that the Kisrawi/Tamara/Naharawan canal system between Samarra and Jarjaraya, 100 km southeast of Baghdad, was completed later under the Sassanian king Chosroes I (531-579 AD) (Adams 1965).

No damming of the Tigris

1.4.2 *Nineveh water supply dams*

Physical evidence of Iraqi dam construction is relatively young and concerns the water supply system built under the Assyrian king Sennacherib (704-681 BC) for the capital city of Nineveh, 355 km north of Baghdad (Fig. 19). Already as a prince he had Urartu and especially its hydraulic works reconnoitred before and after his father's successful campaign against it in 714 BC – an early case of 'industrial espionage'! Although the basic principle, i.e. the integral use of the available rivers is the same, the Assyrian dams were quite

Nineveh's water supply

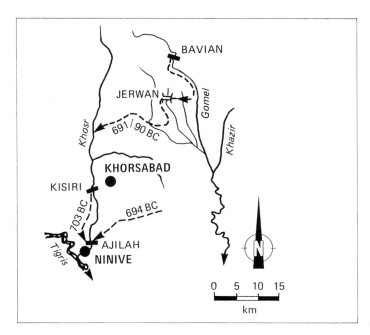

Figure 19. Water supply system of Ninive in northern Iraq (after Jacobsen & Lloyd 1935).

Figure 20. Uppermost part of the Ajilah dam in Ninive; trapezoidal cross section visible at left (photo by the author).

different and novel. In fact according to known details on two of them (Ajilah and Kisiri) they were low diversion weirs of rubble masonry (Fig. 20). Their upstream faces were vertical and in the case of the Ajilah dam lined with ashlar masonry. Broad overflow crests led to the stepped, i.e. inclined, downstream faces. The weirs crossed the rivers diagonally in form of an elongated S in order to obtain the overflow length required by the high flood discharges of the torrential river Khosr. Thus the Ajilah dam had an overall length of some 230 m and a height of about 3 m (Thompson & Hutchinson 1929).

Details of the diversion dams

The first of Sennachherib's dams was completed at Kisiri (or Qayin) in 703 BC to divert water from the Khosr into a 15 km long canal to Nineveh (Fig. 19) (Jacobsen & Lloyd 1935). The second dam built in 694 BC at Ajilah next to the city is located just below the confluence of a canal from the northeastern mountains with the Khosr. It thus formed a sort of terminal basin for the canal and protected Sennacherib's artificial downstream swamp, which was inspired by his campaigns in the south of Iraq in 702 and 700 BC. Unfortunately no details are known about the third dam at Bavian on the Gomel river intended to divert its water into a 55 km long canal to the Khosr (Fig. 19). It was built in the years 691 and 690 BC and included a 300 m long aqueduct wall across the valley of Jerwan measuring 10 m in height, that had at times been mistaken for a dam (Bachmann 1927).

Sequence of the dams' construction

1.5 SOUTHWESTASIAN DESERTS

1.5.1 *The oldest dams in the world*

The world's most ancient, well researched and documented dams to this day were located some 100 km northeast of the Jordan

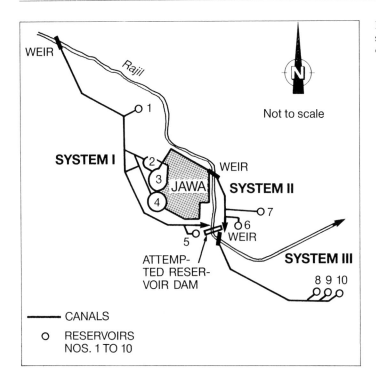

Figure 21. Schema of the water supply systems for Jawa in Jordan (after Helms 1981).

capital Amman (Helms 1981).* They were part of an elaborate water supply system for the town of Jawa, which experienced a brief heyday around 3000 BC (Fig. 21). Located in an inhospitable basalt desert, the city, laid out for some 2000 inhabitants, must have been built very rapidly. It was even established, that its construction had to start together with that of the water system I at the beginning of a rainy season, if the builders and their herds should survive the following dry period.

Oldest dams in the world

From this follows, that the town of Jawa was not developed by local nomads, but was built from scratch by a highly organized group of urban refugees. Like the Israelites under Moses (around 1250 BC) some 18 centuries later, they probably came from the north in search of a new (temporary ?) home. The selected site of Jawa was the most favourable from the points of view of defence and water supply. In the north and east it was confined by the 20 m deep canyon of the Rajil, which has an average yearly discharge of some 2 million m³. Towards south and west the city hill descended to a shallower, broader valley, in which the runoff from the slopes collected in natural pools.

Urban refugees as builders

*There are claims, whereby still older dams were found in Armenia, but the relevant data are nebulous and inconsistent (Agakhanian 1985).

Diversion and reservoir
systems

From these pools the Jawaites satisfied their water needs during construction of the city and they dammed them immediately to form artificial reservoirs (Fig. 21). In order to fill these a 3 km long canal was built from a diversion weir on the Rajil to the northwest of the town. It must have looked like those Sennacherib erected later for Nineveh (see above) and was 150 to 200 m long. The five reservoirs of system I were able to store some 46 000 m^3 or over 0.15 m^3 per inhabitant and his cattle per day during the dry season. Half of the storage capacity was concentrated in Reservoir No. 4, which was impounded by a 4.5 m high and 80 m long dam (Fig. 22).

Figure 22. Ruins of the dam impounding reservoir no. 4 at Jawa; upstream is to the left (photo A. Vogel, Vienna).

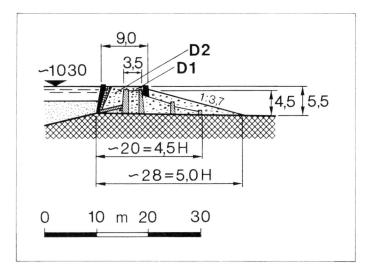

The internal structure of this dam was quite complex (Fig. 23). The water retaining element consisted of two dry masonry walls, which enclosed an earth core of a thickness of 2 m. Additionally an impervious blanket was provided in front of the upstream heel. The stability of the structure was assured by a downstream embankment. The heightening of the dam by 1 m (D2 in Fig. 23) followed essentially the same principles, although the width of the earth core was now increased to some 7 m. A pervious stonefill was provided behind its upstream wall for easy drainage during emptying of the reservoir. Thus the wall was protected against the dangers of water-backpressures, a precautionary measure that was reinvented only in modern times! The raising of the dam as well as a similar reservoir structure attempted across the Rajil itself were never completed because the miraculous desert city collapsed within one generation as swiftly as it arose. It might even have become a victim of its success, in that too many people settled there with the result that they soon overstressed the water supply systems.

Details of main dam

System collapse

1.5.2 *The water supply of Jerusalem*

The Canaanitic city of Urusalim was in a similar situation as Jawa. It expanded in the second millennium BC on a rocky ridge between tow ravines. Its water supply relied on the Gihon spring in the eastern ravine. After the Israelites conquered the city around 1000 BC, their king Solomon (968-928 BC) diverted the waters of the spring by means of a canal into an artificial reservoir near the mouth of the Tyropoeon secondary ravine some 400 m to the south, called 'Pool of Shiloah (or Siloam)' or, in Arabic, 'Birket el

Jerusalem's early water supply

Probatica dam

Hamra' (Wilkinson 1978). Its dam was covered in the second century BC by a massive, new structure about 150 m long.

By contrast part of the Probatica (also Bethesda or Sheep) dam built for the supply of water under king Achaz (741–725 BC) in the Rephaïm valley north of Solomon's temple is still visible (Fig. 24) (Pierre & Rousée 1981). It was a masonry wall up to 13 m high and 40 m long. Its thickness varied only between 6 m at the crest and 7 m at the base. This was barely sufficient for its stability according to modern criteria (including uplift forces). Near the upstream face there was a vertical shaft with an interior section of 1 by 1 m and small inlets at every 2 m of height. These were normally plugged and opened successively when the reservoir was emptied. Such an

Figure 24. Part of the down-stream face of the Probatica dam in Jerusalem; on the right one of the foundation arches for a Byzantine basilica (photo by the author).

ingenious device for the control of the water releases without gates was to reappear again as late as in Yemenite and Roman dams (see below)! From the base of the shaft the water flowed through a gallery to the downstream toe of the dam. Around 150 BC the floor of the gallery was lowered some 2 m. The released water was led into a conduit below the floor of an additional reservoir excavated in the rock downstream of the dam. After construction of Herod Agrippa's I (41-44 AD) city wall across the reservoir, considerable transformations took place also on the upstream side of the Probatica dam.

Menaced by the Assyrian king Sennacherib (see above) the king of Judah, Hezekia (725-697 BC), replaced Solomon's open-air diversion canal from the Gihon spring in 704-701 BC by his famous 533 m long tunnel (Amiran 1976). He also had a new terminal reservoir built further up the Tyropoeon ravine within the city walls (Wilkinson 1978). It later underwent numerous transformations and is today named 'Birkeh Silwan' or, erroneously, 'Pool of Siloam'.

Likewise, three reservoirs 11 km southwest of Jerusalem were later falsely attributed to king Solomon, who in the second of his psalms (Bible) was not referring to them but to his 'Shiloah Pool'. While the lower of these so-called 'Solomon Pools' dates from the 16th/17th century AD, the other two were built towards the end of the first century BC (Mazar 1984). The upper one contained about 85 000 m^3 of water, which were impounded by a retaining wall backed-up by an embankment. The dam in the middle was a gravity wall some 12 m high and 80 m long (Fig. 25). It was able to store about 90 000 m^3. The reservoirs collected the water from nearby springs, which had already been brought to the Temple Mount in Jerusalem by means of a small, 21 km long canal under one of the last Hasmonean kings (167-37 BC). During Herod the Great's (37-4 BC) rule additional water was diverted into the upper reservoir from the El Biyar valley southwest of it. The 5 km long aqueduct included a tunnel with the extraordinary length of 3 km. Under the Roman governor Pontius Pilate (26-36 AD) water from the Arrub valley in the south was brought to the middle reservoir by a 40 km long canal, so that the system was now capable of a daily delivery of some 1400 m^3 of water to the Temple area. It operated continuously until late in the 19th century and was restored in 1924.

The abovementioned Hasmonean rulers also built a 9 km long water supply canal to an estate they established in the Jordan valley near Jericho, some 20 km northeast of Jerusalem, to grow date-palms. It also served as their residence during the mild rainy season (Garbrecht & Netzer 1991). The water intake in the Qelt ravine to the west included a 5 m high masonry wall. A smaller weir had

Figure 25. Downstream view of the dam impounding the middle one of the so called 'Solomon Pools' near Jerusalem (photo by the author).

Smaller Hasmonean dams

already been used in the second century BC at the nearby fort of Dok (or Dagon) on Mount Karantal to bring water by means of a short canal to the cisterns of the fortress (Garbrecht & Peleg 1989). A larger weir was built around 100 BC near the fort of Hyrkania, 17 km east of Jersualem, because the remains of its foundations are 17.5 m long and 4 to 5 m wide.

1.5.3 Irrigation dams in Yemen

Irrigation works at Marib

A famous event during the reign of the abovementioned Solomon was the splendid visit of queen Bilqis from Saba in Yemen, over 2000 km southeast of Jerusalem (Bible). Marib was the capital city of Saba, 120 km east of Sana. Its wealth based on more than a millenium of irrigation and on the traffic of frankincense and myrrh produced near the southern coast of Arabia. These products were much coveted in the Mediterranean world. Since about 1500 BC the Sabeans completely dammed their main river Danah (Hehmeyer 1991). Only traces of these early structures were found

Figure 26. Upstream view of the southern outlet of Marib dam in Yemen; on the left the abutment wall for the embankment (photo U. Brunner, Bassersdorf/CH).

immediately downstream of the southern outlet of the great dam. Its construction began around 510 BC (Fig. 26).

It is unknown when the great dam reached its final configuration (Brunner 1983). From what is visible or may be retraced today its main part consisted of an embankment about 700 m long and up to 20 m high (Fig. 27). It had rather steep slopes of 1:1.8 on both sides and no crest road. The fill was placed in layers parallel to the slopes and not in horizontal layers as usual. This facilitated subsequent heightenings and repairs. The upstream slope was partly faced with a 0.2 to 0.4 m thick layer of stones. There was no special impermeable element in the dam.

Configuration of great dam

On the rocky river banks at both ends of the embankment impressive outlet structures built of excellent ashlar masonry were located. The two outlets, one in the north and one in the south, had their sills at almost the same elevation, 13 m above the river bed, and they were partly provided with gate slots in the northern structure. This also contained a 50 m long overflow spillway to the north, the sill of which lay some 3 m above those of the outlets. Up to it the original reservoir capacity was some 30 million m^3 or barely 15% of the average annual river discharge. Because the outlets were not placed much lower the reservoir silted up rapidly and much silt entered into the canals. This, however, was desirable because the silt acted as fertilizer although it heightened the fields by 0.01 m each year. A constant adaptation of the irrigation system was therefore necessary. This also proves that the dam rather was intended to lift and divert the river's flow and not so much to store it.

Impressive and sophisticated outlets

Between the spillway sill and the 4 m higher dam crest there was a flood storage capacity of an additional 30 million m^3. This together with the discharge capacity of the spillway and the outlets of

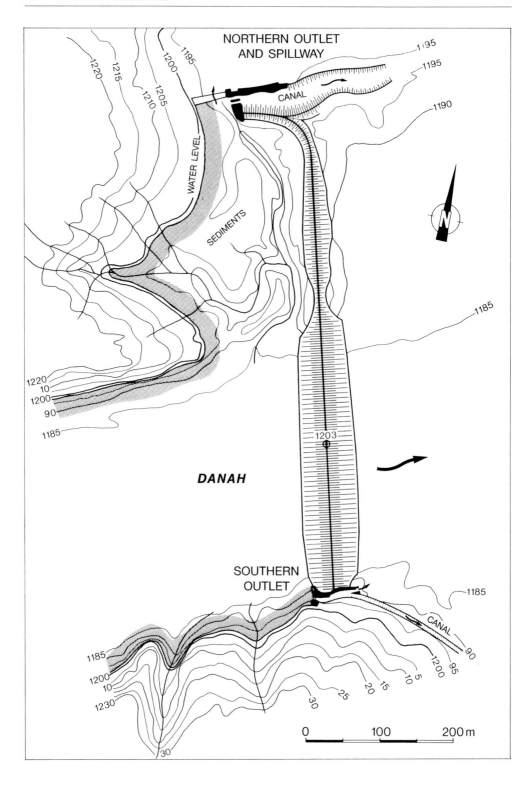

some 1500 m^3/s, was enough only to handle relatively frequent floods. Consequently the dam was breached quite often by extreme floods, which occurred in the last two and a half centuries of its existence, about once every 50 years. A special event were the devastations by the Romans in 25–24 BC during their unsuccessful campaign against Marib. After some 1300 years of service, the final failure early in the seventh century AD was recorded in Mohammed's (570-633 AD) Koran with the following words: 'Eat ye (the Sabeans) of your Lord's supplies and give thanks to Him: goodly is the country and gracious is the Lord. But they turned aside : so we sent upon them the flood of Iram (name of place or tribe) and we changed their gardens into two gardens of bitter fruit and tamarisk and some few jujube trees' (34th surah). This time in fact the dam was not repaired and most of nearly 50 000 people, whose livelihood depended on it, emigrated. In the years 1984-1986 a modern embankment dam was built some 3 km upstream of the ancient one. It has a storage capacity of 400 million m^3.

There are quite a few other ruins of ancient dams in Yemen, of which the majority has so far neither been studied in detail nor been dated. An exception is the Ghayl (or Amir) dam 85 km northwest of Marib (Robin et al. 1981). Its initial construction in the fourth/third century BC consisted of a simple masonry wall, which must have failed shortly after completion. The reconstruction in the second century BC included a second wall upstream, whereby the 6.4 m wide intermediate space was filled with earth. This type of structure was already in use by the Mycenaeans and the Urartians (see above) and also adopted by the Sabeans for some of their smaller dams around Sana in secondary ravines to retain part of the flood runoff (Table 2) (Siewert 1979, 1982).

As can be seen from Table 2 the majority of the dams were, however, of the gravity type that had also been used by the Mycenaeans and, later on, by the Assyrians and the Israelites (see

See Page 26

Figure 27. Plan of the Marib dam in its final configuration (topography Hunting Surveys Ltd.).

Final failure of great dam

Other Sabean dams

Table 2. Sabean dams near Sana.

Name of ravine		Height (m)	Length (m)	Base width (m)	Ratio width/ height	Characteristics
Dodan	Lower[1]	(26)	60	28	(1.1)	Two earth cores
	Middle[1]	?	20–30	18	?	Gravity/imperm. ?
	Upper	10	65	18	1.8	Gravity/plaster
	Gailan	6	32	4.5	0.8	Gravity/imperm. ?
	Sirwah	5–6	20–30	4–5	0.8	Earth core
	Tanam	4.5	20	?	?	?
	Wurda	7–9	70	10	1.3	Gravity/plaster

[1]Second structures.

Himyar dams

above). They consisted of two dry masonry walls with rubble filling in between. Sometimes both walls were inclined as much as 50% (horizontal:vertical) (Upper Dodan) and the upstream one was made impermeable by plastering. The dams identified around Yarim, 120 km south of Sana, near Zafâr, capital city of the kingdom of Himyar, were exclusively of this gravity type. Their reign in the first century BC became the most powerful in Yemen (Table 3) (Siewert 1979, 1982).

The most interesting one of these structures is the Asid (or Adraa) dam, 100 km south of Sana, because a good part of it is still standing (Fig. 28). Each of its two outer walls was stepped back five

Table 3. Himyar dams near Yarim.

Name(s) of ravine	Height (m)	Length (m)	Base width (m)	Ration width/ height	Characteristics
Ans (Kawla)	3	100	?	?	Gravity/plaster
Asid (Adraa)	19	60	20	1.1	Gravity/plaster
Djobar (Jubar)	11	100	16	1.5	Gravity/plaster
Taiy as–Sabisa	~4	80+60	7	1.8	Gravity/imperm. ?

Figure 28. Ruins of the upstream face and the outlet tower of Asid dam in Yemen (photo H. Marti, Buchberg/ZH).

times to obtain an average inclination of about 30%. In the centre of the rubble fill there was a 2.3 m thick, vertical sand drain over almost the entire height, which was enclosed by dry masonry. Its purpose was to keep the downstream half of the dam dry and free of uplift forces in case of leakage through the plaster on the upstream face. This device is encountered again in modern dams, primarily in embankments! Also the separate outlet tower with water intakes at various levels akin to the Probatica dam in Jerusalem (see above) may be considered as very advanced. In a poem around 400 AD king Abûkarib Ascâd boasted that there were 80 irrigation dams in the area of Zafâr (Wissmann 1966).

Details of Asid dam

1.5.4 *Nabatean dams*

Another people who gained from the frankincense and myrrh trade between southern Arabia and the Mediterranean world were the Nabateans. Of uncertain but probably South Arabian origin they installed themselves, like the Jawaites and Canaanites before them, in the sparsely populated desert area in the south of Israel and Jordan near the northern end of this trading route. In the second century BC the Nabateans erected a kingdom with Petra, 185 km south of Amman, as their sumptuous capital. Located in a desert ravine it could only survive on the basis of extensive and elaborate hydraulic works (Lindner 1987). Of great importance was the control of the flash floods in the ravine. By means of a 14 m high and 43 m long rockfill dam some 2 km southwest of the city part of the flood runoff was diverted through a 400 m long tunnel into a secondary valley, where its impetus could dissipate. After a flash flood in 1963 drowned 24 visitors of the ancient city the old dam was reconstructed (Fig. 29) (Time 1963).

Nabatean kingdom

Flood protection for Petra

Figure 29. Reconstructed flood diversion dam upstream of Petra in Jordan and, on the right, entrance to the Nabataean diversion tunnel (photo A. Vogel, Vienna).

Check dams for water
and soil conservation

In the abovementioned secondary valley – and in many other, smaller ravines – the Nabateans built flood control or check dams, which very often served also to retain the eroded sediments for cultivation on them (Lindner 1987). They actually became great masters in water and soil conservation. In an area of 130 km^2 around Avdat, 115 km south of Jerusalem, 17 000 check dams or 130 per km^2 were found (Kedar 1957). Their silted up reservoirs increased the area of arable land from 0.5 to 7.5 km^2, i.e. fifteen-fold. On average these check dams were 1.8 to 2.0 m high, 40 to 50 m long and 2.0 to 2.5 m wide. Like some of the Mycenaean, Urartian and Yemenite dams described before they consisted of two stepped back walls. The space between them was filled with rubble and earth. Every dam had a spillway for the passage of the runoff not absorbed by its sediments.

Few Nabatean storage
dams

For the storage of water the Nabateans preferred underground cisterns, probably because they kept the water cool and minimized losses due to evaporation. Only rarely did the Nabateans resort to storage dams. The most important such instance is substantiated near Kurnub, 85 km south of Jerusalem (Woolley & Lawrence 1914-1915). There, three dams were built close together across a ravine at the foot of the city to store some 10 000 m^3 of water. Largest of the three dams was the lowest along the river course: 11 m high, 24 m long and 7.8 m wide at the crest (Fig. 30). The middle and upper dams are silted up so thoroughly that only their crest dimensions are known: length 20 and 53 m, width 5.0 and 3.4 m respectively. All were built with masonry set in lime mortar and the lowest one was kept in good condition so as to serve to this very day.

Details of the Kurnub
dams

Smaller Nabatean storage dams were found near Petra and at

Figure 30. Downstream view of the largest of the three storage dams near Kurnub in Israel (photo T.Tsuk, Ramat Gan/ Israel).

Sabra some 7 km southwest (Lindner 1987). In a lateral ravine about 100 m above the Sabra valley there is a 4.6 m high, 14 m long and 4 m wide wall of ashlar masonry carefully set in mortar, which could store about 3600 m³ of water. From the outlet measuring 0.7 by 0.7 m near the base of the dam the water was brought down to the valley floor by a very steep, tortuous canal. The reservoir near Petra was part of the final section of the northern Hubta aqueduct to the city from the east. Located in a narrow gully it contained only 300 m³ of water. Approximately 40 km south of Petra near Humayma there is a silted up but splendidly preserved dam, 3.7 m high, 11 m long and 4.4 m wide, built like the abovementioned Sabra dam (Oleson 1987). Smaller Nabatean dams

The Nabatean kingdom attained its greatest extension in the first half of the first century BC, when it reached as far north as Damascus in Syria for some years. But here it also collided with the Romans, under whose predomination it remained a separate state until it was annexed in 106 AD. The refined water and soil conservation measures were retained under the rule of the Romans who applied the technique in other parts of their empire as well (see below). Adoption of Nabatean techniques by Romans

1.5.5 *Check dams in Baluchistan*

One region where water and soil conservation in the Nabatean manner was practiced long before their appearance on the world scene, is Baluchistan on the southeastern edge of the Iranian desert (today in Iran and Pakistan) (Raikes 1964-1965). In the east the region borders the arid Indus valley, where a civilization based on sophisticated hydraulic works (no dams) flourished between 2500 and 1800 BC. Possibly the technique of check dams in Baluchistan, which are locally called 'Gabarbands' or 'Dams of the Zoroastrians' (Iranian religion of the seventh century BC) was even more ancient. Very old water and soil conservation dams

Like the Nabatean check dams many of those in Baluchistan consisted of two stepped back dry masonry walls with an intermediate earth and rubble filling. Another type had only one solid masonry wall, sometimes reinforced by buttresses on both faces. Blocks weighing as much as 2 tons were used and the upstream heels of the dams were protected by aprons of rubble. Most of the dams were built to retain sediments for cultivation on them. Only the larger ones of some 20 m in height were intended for the storage of water, as e.g. those on the Hab and Turkabar rivers, about 250 km north of Karachi in Pakistan. Today, possibly 5000 years after its introduction, the technique is still in use. Few Baluchi storage dams

1.6 SRI LANKA

Natural conditions on
Sri Lanka

The island of Sri Lanka (formerly Ceylon) at the southern tip of
India covers an area of 66 000 km² (Fig. 31). While its mountainous southwestern quarter receives ample rainfall, the dryer rest
of the island is better suited for agriculture. The average annual
precipitation of 1 to 2 m is still considerable although it concen-
trates on the three months of December through February. Moreover the thin top soil cover mostly rests on a compact subsoil of

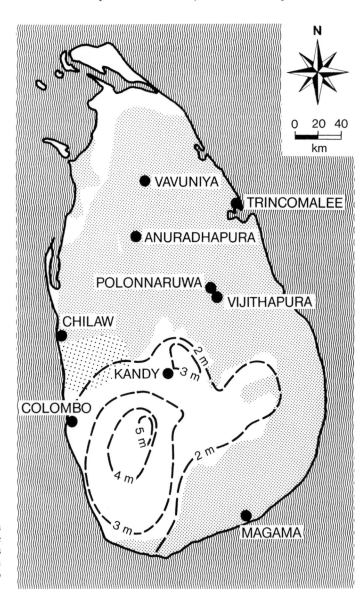

Figure 31. Map of Sri Lanka
with curves of equal average
yearly precipitation in metres
and the (hatched) areas with
large reservoirs (after Fernando
1982).

crystalline rock. Therefore its absorption and retention capacity is very limited and the concentrated rainfall runs off rapidly (Fernando 1982). The obvious answer to such conditions, is artificial water storage. The first farming communities in northern and eastern Sri Lanka already built small and primitive village reservoirs in great numbers.

1.6.1 *Large Singhalese reservoirs*

In 544-543 BC the adventurous Indo-European prince Vijaya of Bengal (eastern India) with 700 of his followers installed the Singhalese dynasty of kings in Sri Lanka. These and/or their governors set out to build large reservoirs that did not replace the abovementioned village ponds, but rather supplemented them. One of the earliest of these reservoirs was that of Panda* on the westcoast. It dated from about 370 BC and stored already 9 million m^3 of water (Table 4) (Parker 1909). The dam axis was not straight but skilfully adapted to the topography, so as to avoid low ground. It also showed already one of the main characteristics of the Singhalese dams, i.e. their large crest length (Fig. 32). With 2.4 m the crest width was still rather narrow and the slopes of 1:2.5 on both sides relatively steep.

Earliest Singhalese dam

The slopes of the Bassawak and Tissa dams were flattened to 1:3 about two generations later. They served the water supply of Anuradhapura in the north of Sri Lanka, which had meanwhile become the capital city for the following 1000 years (Fig. 33 and Table 4). Again the axes of both dams were not straight. They bulged downstream to maximize the reservoir capacities. This, naturally, increased the volume of the embankments, but in those days construction was hardly accustomed to present day speed, so that the number of forced labour needed could be kept within reasonable limits. Under the assumption of a performance of 0.5 m^3 per person and day and of 200 working days per year (deducting the rainy season) the 620 000 m^3 of fill for the Tissa dam thus required altogether some 6000 man-years. In other words for a construction period of six years, 1000 labourers were therefore needed. As many people are reported to have been permanently employed just to sweep the streets of Anuradhapura!

Two dams for Anuradhapura

Construction of the Tissa dam

Not much later, similar reservoirs were built for the water supply of the vanished secondary capital city of Magama on the southcoast (Fig. 34 and Table 4). The oldest dam – also named after

*The frequent suffix 'wewa' (Singhalese) or 'kulam' (Tamil) is omitted here, because it also means reservoir. The same applies for the names of rivers (= 'ganga' and 'oya' or 'aru' respectively).

Table 4. Ancient Singhalese embankments on Sri Lanka.

Period of con-struction	Name(s)	Location (Fig. or km from city)	Height (m)	Length (km)	Capacity (million m³)
370 BC	Panda	26E Chilaw	7	*2.6*	*9*
310 BC	Bassawak (Abhaya)	Fig. 33, center	7(10)	1.8(1.2)	4(2)
300 BC	Tissa/Anuradhapura	Fig. 33, center	*8* (11)	*3.3*(3.5)	(3)
300 BC	Tissa/Magama	Fig. 34, center	6	1.2	4
300 BC	Paskanda Ulpotha I	23 SE Polonnaruwa	*17*	?	?
3rd c. BC	Sangilikanadarawa	Fig. 33, top right	5	2.5	8
Late 3rd c. BC	Yoda (Duratissa)	Fig. 34, bottom	4	1.0	*10*
3rd/2nd c. BC	Pavat	Fig. 33, top	9(14)	3.0(3.1)	*22*(33)
Early 2nd c. BC	Vavuni (Peliwapi)	40 NW Vavuniya	7 (14)	*4.1*(3.3)	17(49)
2nd c. BC	Yoda Kandiya	Fig. 34, top	5	3.5	11
2nd/1st c. BC	Batalagoda	33 NW Kandy	9	1.8	4
80 BC	Nuwara	Fig. 33, center	11(11)	*4.8*(6.8)	*43*(44)
1st c. AD	Wahalkada	37 E Vavuniya	*(19)*	*(5.1)*	*(53)*
70+290 AD	Minneriya	15 NW Polonnaruwa	*(21)*	(1.9)	*(136)*
290 AD	Hurulu	26 NW Polonnaruwa	*(25)*	(2.4)	(68)
290 AD	Kaudulla	24 N Polonnaruwa	(15)	*(9.2)*	(51)
460 AD	Paskanda Ulpotha III	23 SE Polonnaruwa	*34*	?	?
40 BC+470 AD	Balalu + Kala	Fig. 33, bottom	(15)	*(9.7)*	*(123)*
540 AD	Nachchaduwa	Fig. 33, center	17(17)	1.7(1.6)	15(56)

In parenthesis: Modern dimensions after reconstruction.
Italics: Record breaking dimensions.

Figure 32. Crests, kilometres long, are one of the main characteristics of Singhalese dams in Sri Lanka; picture shows the Nachchaduwa dam (photo by the author).

See Page 35

Figure 33. Map of the reservoirs and diversion canals (broken lines) around Anuradhapura; figures in frames indicate reservoir capacity in millions of m³ (for reconstructed reservoirs, present values) (after Brohier 1934-1935).

king Devanampiya Tissa (307-267 BC) – had a markedly larger crest width (5.3 m) as well as unusually flat slopes, namely 1:5.1 on the upstream and 1:4.4 on the downstream face. Particularly noteworthy is the fact that the reservoir received water by means of a canal 4 km long from the Kirindi river, which flowed through the city. The 4.5 m high diversion weir of large blocks crossed the river not perpendicularly, but at an angle of about 45°. Towards the end

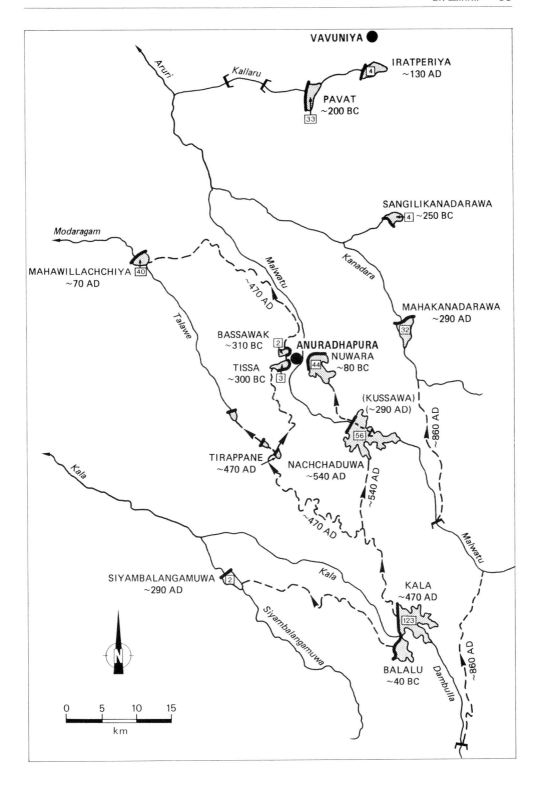

VAVUNIYA ●

IRATPERIYA
~130 AD

Aruri

Kallaru

PAVAT
~200 BC
33

SANGILIKANADARAWA
4 ~250 BC

Kanadara

Modaragam

MAHAWILLACHCHIYA 40
~70 AD

Malwatu

~470 AD

MAHAKANADARAWA
~290 AD
32

Talawe

BASSAWAK
~310 BC
2

ANURADHAPURA

NUWARA
44 ~80 BC

TISSA
~300 BC
3

(KUSSAWA)
(~290 AD)

56

~860 AD

Kala

TIRAPPANE
~470 AD

NACHCHADUWA
~540 AD

~540 AD

~470 AD

Malwatu

SIYAMBALANGAMUWA 2
~290 AD

Kala

KALA
~470 AD

Siyambalangamuwa

KALA
123

N

BALALU
~40 BC

Dambulla

~860 AD

0   5   10   15

km

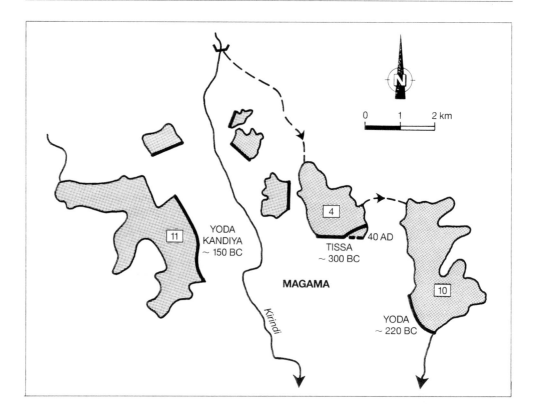

Figure 34. Map of the reservoirs and diversion canals around Magama (after Parker 1909).

of the third century BC the spillway discharge of the Tissa dam was used to fill the additional Yoda (or Duratissa) reservoir. Thus for the first time a system of dams and canals permitting the best possible use of the available water resources had been created. In this respect the Singhalese were to achieve a hitherto unheard-of mastery!

1.6.2 Remarkable structural details

Slopes of Singhalese dams

With the exception of the masonry diversion and overflow weirs, all the ancient dams in Sri Lanka were homogeneous embankments. This was adequate in view of their moderate height, but required relatively flat slopes (Fig. 35). It remains unclear whether the successive flattening of the slopes was continued. The Batalagoda and Nuwara dams of the beginning of the first century BC again had slopes similar to the very old Panda dam (Table 4) (Parker 1909). The Paskanda Ulpotha dam, which after its second heightening in 460 AD reached the record height of 34 m (exceeded only in 1675 by the St. Ferréol dam in France), had a pronouncedly steep 1:2 upstream slope (Fig. 35) (Benson et al. 1983). Incidentally the upstream faces of most Singhalese dams were protected by a layer of cobblestones against washing-out under wave action (Parker 1909).

Upstream wave protection

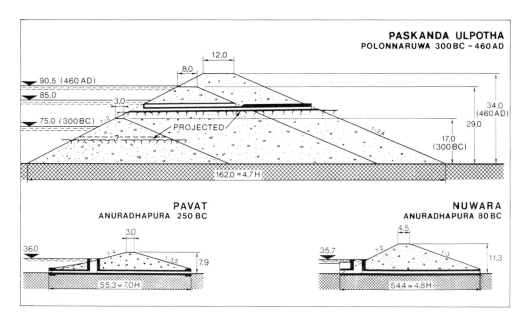

Figure 35. Cross sections of three ancient dams in Sri Lanka (after Parker 1909 and Benson et al. 1983).

Figure 36. Part of the ancient overflow wall at the Balalu-Kala reservoir (left); background shows the modern spillway (photo by the author).

Besides, all dams were protected against over-topping by means of one or several spillways over rocky depressions along the reservoir rim. The latter were sometimes expanded as needed (Parker 1909), and where this was unfeasible, thick overflow walls of dry masonry were built (Fig. 36). They were made impervious by upstream wedges of bricks laid in lime mortar (Fig. 37). Even at the Nachchaduwa spillway this provision was maintained although the stones of the wall had been laid in mortar. Its downstream face was carefully built of large blocks whereby the uppermost were

Elaborate spillways

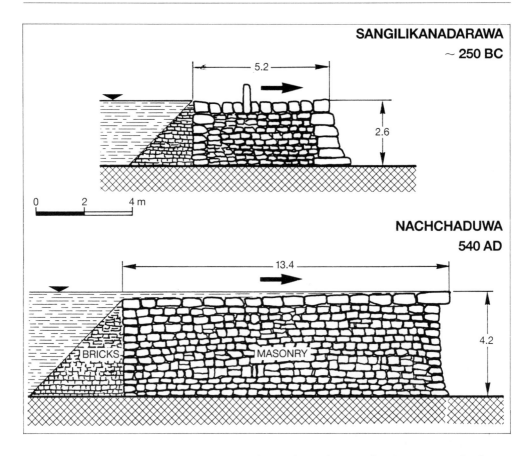

Figure 37. Cross sections of overflow walls at two Singhalese dams; the upper was 82 m long, the lower one 51 m (after Parker 1909).

Sophisticated outlet works

projecting in order to direct the overflowing water as far downstream from the wall's toe as possible. Along the crest of the Sangilikanadawara spillway, and others, stone posts were provided at 3 to 5 m intervals. This allowed to raise the storage water level by means of planks or trunks after passage of the last flood. This was a rather dangerous practice since there is no knowing as to which has been the last flood of a season.

An even better testimony of ancient Sri Lanka's high standards of hydraulic engineering than the one of the described spillways is provided by the outlet works incorporated in all dams from the beginning, often at various elevations. Earlier, analogous arrangements are only known to have existed in Anatolian dams (see above). As can be seen in the cross sections of the Pavat and Nuwara dams in Figure 35 the central part of the Singhalese outlets was a vertical shaft in the upstream face of the dams (Parker 1909). On the average it had a clear width of some 4 m parallel to the dam axis and of about 3 m perpendicularly to it, i.e. in the direction of the flow of water (Fig. 38). The water was brought from the reservoir to the shaft by means of one or two covered, rectangular

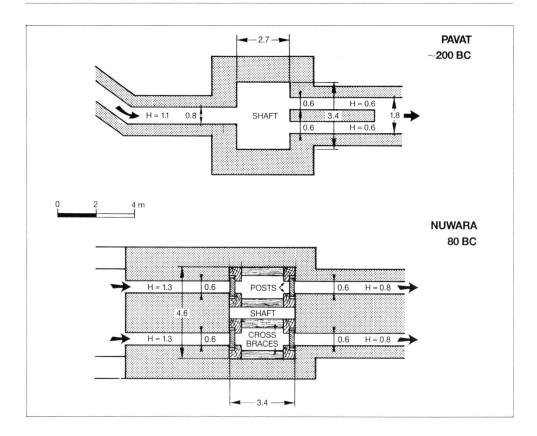

conduits with a total clear cross section of about 1 m². Sometimes there was a short canal ahead of the conduits, which for example was 22 m long and 4.5 m wide at the Nuwara outlet. In most cases the water flowed from the shaft to the downstream toe of the dam through two conduits, the total clear cross section of which was smaller or equal to that of the upstream conduit(s).

The walls of the shafts and conduits consisted mainly of brick masonry lined on the water faces by up to 4 m long, 0.6 to 0.9 m high and 0.25 to 0.30 m thick precisely fitting stone slabs (Fig. 39). Similar or thicker slabs were used for the floors of the shafts and the conduits as well as for the latter's covers. All these structures were enclosed by a thick layer of fat loam to prevent seepage and subsequent erosion out of or along them, all too often with disastrous damages to embankments. As no slots for gates were found in any of the stone revetments of the shafts, it must be assumed, that these were supported by wooden structures, as shown hypothetically in Figure 38 (bottom) (Bligh 1910). The wooden gates were probably operated by levers.

Figure 38. Plans of outlet shafts at two Singhalese dams showing stone lined brick structures and the hypothetical wooden supports (bottom) for the wooden gates; H = clear height (after Parker 1909 and Bligh 1910).

Structure of the outlets

Figure 39. Reconstructed part of the ancient outlet at Balalu-Kala reservoir (photo by the author).

1.6.3 *Water resources management*

Systematic development of water resources

As mentioned above, the Singhalese became masters in the systematic development of water resources. Unfortunately it is hardly known which parts of their systems were in operation simultaneously. It appears as certain, that they never developed a centralized hydraulic bureaucracy as did the ancient Egyptians or Chinese (Leach 1959). The earliest system of canals and reservoirs near Magama has already been described. At some unknown date the 43 million m^3 reservoir of Nuwara – built about 80 BC near Anuradhapura – was provided with a 22 km long adduction from the Malwatu River upstream of the city (Fig. 33). Under king Mahasen (275-301 AD) the intake disappeared in the Kussawa reservoir, which was in turn submerged 250 years later by the 56 million m^3 reservoir of Nachchaduwa.

Anuradhapura system

In similar fashion, king Dhatusena (459-477 AD) had the pre-Christian reservoir of Balalu integrated into the 123 million m^3 reservoir of Kala, some 40 km southeast of Anuradhapura (Fig. 33 and Table 4). The most important hydraulic enterprise of Dhatusena however, was the construction of a 87 km long and 12 m wide canal from his new Balalu-Kala reservoir to those of Bassawak and Tissa, near Anuradhapura. Later the canal was prolonged to the then already existing Mahawilachchiya reservoir in the Talawe valley, which was moreover fed by a branch of Dhatusena's main canal (Fig. 33). Over its first 30 km of length the latter had a slope of only 1:10 000, which was very demanding not only for the surveyors but also for the builders.

High precision work

1.7 CHINA

In view of the humid, subtropical climate irrigation had no priority in China. Instead it had to share its significance with an increasing struggle for river embanking and training as well as an early and constant preoccupation with inland water transport (Needham 1971). Moreover, great reservoirs were not the most characteristic form of the sophisticated Chinese hydraulic technique, although the still existing Anfengtang or Shao reservoir dates back to before the 'Warring States Period' (402-221 BC). The same holds true for the first diversion weirs and irrigation or navigation canals. Many more such works were undertaken after the unification of China by emperor Cheng in 221 BC and during the ensuing 500 years, until the time when China fell apart again (317-589 AD, era in which the western half of the Roman empire also disintegrated).

Earliest dams in China

1.7.1 *Storage reservoirs*

From this ancient period of China seven reservoirs are recorded, which were all located in the centre of the country, west of and around Shanghai (Fig. 40) (Zheng Liandi 1991). The oldest is, as already mentioned, the Afengtang reservoir built from 589 to 581 under the supervision of Sun Shuao, a minister to a duke of king Ting (606-586 BC). The dam rested on a foundation of sand and gravel and consisted of alternating layers of straw and earth, into which a line of chestnut piles was driven upon completion of the embankment. The latter closed off a plain between two tributaries to the Huai river towards the north and west. It is still in operation today and impounds 100 million m^3 of irrigation water (Fig. 41).

Huge Afengtang storage

A similar reservoir flooding a plain was that of Hongxi, built after the 'Warring States Period' around 100 BC and also located near the Huai river. It was later complemented by more than a dozen smaller reservoirs (Fig. 42). Due to the increase in population, it was thereafter no longer possible to build reservoirs in the plains and Hongxi as well as others (Jian, Aijin and Lian described below) later were even drained. The next reservoirs were thereafter built mostly in the hills. This required shorter dams. One of these was the Maren earth dam, constructed from 48-32 BC under the direction of Zhao Xincheng, president of the respective county. It was 16 m high and 820 m long and consisted of clay throughout. It had masonry outlets as well as two spillways. The dam was reconstructed in 1958.

From plain to hill reservoirs

The Jian (or Jin) and the Aijin (or Chengongtang) reservoirs at the foot of the hills whose runoff they collected were built around 140 and 190 AD respectively. The first one required a 65 km long dam with a height of some 4 m and 69 outlets. The latter were

Extremely long Jian embankment

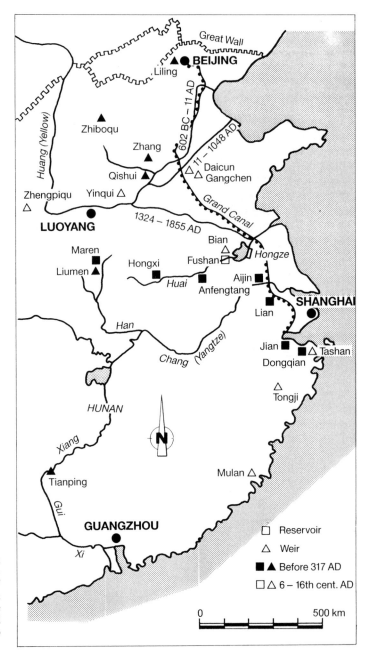

Figure 40. Map of eastern China, with location of reservoirs and weirs built before 317 AD and from the 6th to the 16th century (various courses of the lower Huang river, Great Wall and Grand Canal according to Needham 1971).

operated not only for irrigation purposes but also to control the floods as much as possible. Similarly the just 2.8 km long Aijin earth dam was provided with an outlet and overflow in the centre as well as with additional spillways at both ends. Originally built for irrigation, it was reconstructed in the thirteenth century to also supply

Figure 41. Upstream face of the Anfengtang storage dam in China (photo by Zheng Liandi, Beijing).

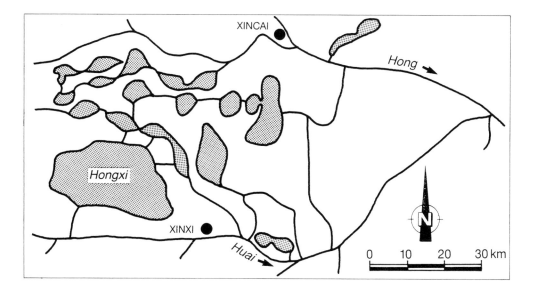

water to the Hangou section of the Grand Canal (navigation, Fig. 40).

The last two typical hill reservoirs were built before the disintegration of China in 317 AD. The still existing Dongqian lake was formed in the third century AD by damming seven outflows or depressions around one valley. At each dam there was an overflow weir, the width of which was either small enough to permit the lifting of boats over it by means of winches, or, alternatively it had slopes on both sides gentle enough to allow for boats to be pushed over it. Four of the seven weirs also had gates for the release of

Figure 42. Map of the Hongxi reservoir and the ones that were added to it up to about 500 AD (after Zheng Liandi 1991).

Last storage reservoirs

Figure 43. Plan of the Lian double reservoir near the city of Danyang (after Zheng Liandi 1991).

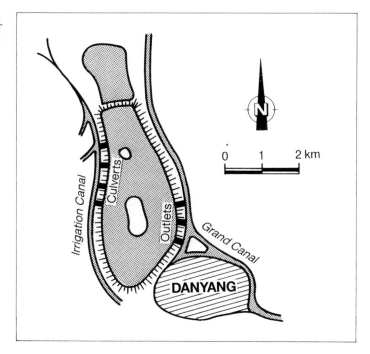

surplus water. Today the Dongqian reservoir can store 40 million m^3 and it is the main water source in the area.

The Lian reservoir was built from 304–306 AD for irrigation and flood control, but after about 600 AD it was also used for the supply of water to the Grand Canal. It actually consisted of two reservoirs, because an intermediate dike subdivided the total height, which had to be dammed (Fig. 43). This is an economical alternative in case of either inability or unwillingness to build high dams. The intermediate dike had three masonry outlets for the release of water into the lower reservoir. The same number of outlets and an overflow weir in the main dam fed the navigation canal to the east of the reservoir, while 12 culverts served the irrigation canal to the west.

Subdivided head in Lian reservoir

1.7.2 *Diversion weirs*

In contrast to the storage reservoirs the recorded diversion weirs were built in the mountainous hinterland of China from north to south (Fig. 40) (Zheng Liandi 1991). The one of Zhiboqu and the twelve on the Zhang river in the north date back to 453 and 422 BC respectively, i.e. just prior to the 'Warring States Period'. Nothing is known about their structural characteristics, except in the case of the Tianping weir built in 219 BC in the south. Its purpose was to raise the water level in the uppermost reach of the

First diversion weirs

Xiang river and to divide its discharge into two navigation canals (Fig. 44). Whereas the North Canal connected the impoundment with the Xiang further downstream, the South Canal led to the Li river, tributary to the Gui and Xi rivers flowing to Guangzhou (formerly Kanton). Since the Xiang flows into the Chang (Yangtze) river, on the estuary of which lies Shanghai, the ingenious scheme created an inland waterway between the two important cities as well as between the southcoast and the central province of Hunan (Fig. 40).

According to its divider role the weir was V-shaped in plan with a longer northern and a shorter southern leg. The total length was 470 m and the height reached 3.0 m above foundation. Under the 2 m thick upstream wall of ashlar masonry, the latter was reinforced by closely spaced wooden piles. Since the whole length of the weir was used to discharge surplus water not needed in the canals, a gently sloping rubble fill was provided on the downstream side (section in Fig. 44). It was covered with tightly packed vertical slabs, which dissipated much more of the energy of the overflowing water than ordinary horizontal slabs at the same time being more stable. Another special feature of the Tianping weir was the 'ploughshare' built in front of its apex, which helped to keep the division of the water flow to the two canals constant at all river discharges.

Ingenious Tianping scheme

Structure of Tianping weir

Figure 44. Plan and cross section of the Tianping diversion weir in southern China (after Zheng Liandi 1991).

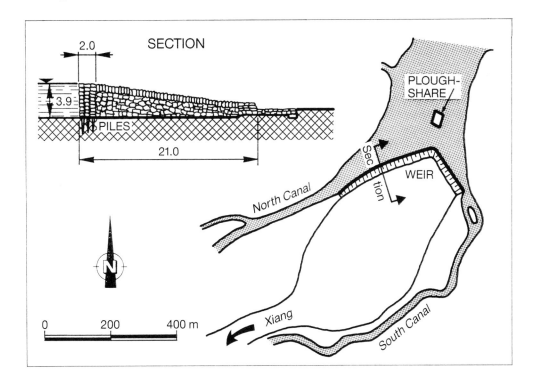

Later weirs

Not much is known about the weirs built after the technological climax of Tianping. In 204 BC one of the rather rare timber weirs was built for a navigation canal at the mouth of the Qishui into the Huang (Yellow) river, which at that time took one of its most northerly courses (Fig. 40). The Liumen dam built in 34 BC was an earthfill, while the Liling weir constructed in 250 AD west of Beijing (formerly Peking) consisted of a timber crib filled with rocks.

1.8 PRECOLUMBIAN MESOAMERICA

1.8.1 *Formative period*

Hydrology of
Mesoamerica

The climate of Mesoamerica (here inclusive of the southwestern USA, Fig. 45) is mostly arid in the central highlands but humid and tropical along the coasts. In both cases however, rainfall and runoff are unevenly distributed over the year. Therefore, the early farming communities, which came into being long before the famous civilizations of the Zapotecs, Toltecs, Maya, Anasazi and Aztecs, already built dams to store water in the rainy seasons for use during the dry periods.

Figure 45. Map of Mesoamerica showing locations of the ancient dam sites.

The oldest such structure found so far, is the Purron earth dam near San José Tilapa at the southern end of the Tehuacan valley,

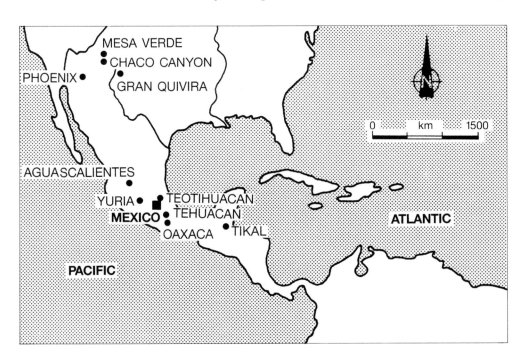

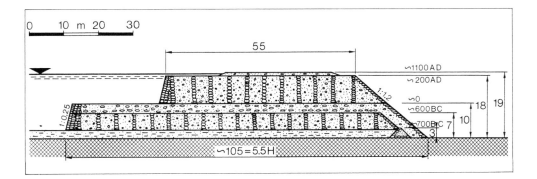

Figure 46. Cross section of the Purron dam in Mexico with its various construction stages (after Woodbury & Neely 1972).

260 km southeast of Mexico-City (Woodbury & Neely 1972). Since the Lencho Diego torrent, which had been impounded, has broken through the dam, its internal structure is now visible (Fig. 46). The first, 3 m high construction stage is dated to around 700 BC. Its modest reservoir was silted up relatively fast so that the dam had to be heightened by 4 m about one century later. The addition consisted of dry masonry cells, which were filled with compacted, sandy soil. On the upstream side there was a strong retaining wall, while the sloping downstream face was covered with stones. The dam was now 400 m long and impounded 1.4 million m^3 of water. The width of its base which rested on the silt deposits behind the first construction stage, was grossly overdesigned. However, this allowed further heightenings with only marginal widening.

Large Purron dam built in stages

Around 200 AD the Purron dam attained a height of 18 m, a volume of 370 000 m^3 and a storage capacity of 5.1 million m^3 (Fig. 46). In all probability the small heightening occurred about 1100 AD rather in connection with a religious construction (pyramid) than a further utilization of the reservoir. But even so the latter extended over more than one millennium! Thereby water was withdrawn through two ditches at the right end of the dam which were closed with trunks, blocks and earth. These ditches also served to spill floods. A short canal connected them with the fields below the dam. Another remarkable detail is the 5 m high dike built upstream of the main dam during its first heightening, intended either as a cofferdam to protect the construction site or to retain the sediments. The dike was later not heightened in parallel to the main dam.

Xoxocotlan diversion weir

A smaller dam from the fifth century BC was found near Xoxocotlan, southwest of Oaxaca or 370 km southeast of Mexico-City (Fig. 45) (Mason et al. 1977). It consisted of two 10 m high, 40 m long and 40 m thick walls of boulders set in lime mortar. The over-flown crest was covered with cut limestone blocks. In plan the two sections of the dam formed a 'V', the apex of which pointed upstream and contained the intake to a canal.

1.8.2 *The high civilizations*

Monte Alban water
supply

However, the irrigation perimeter fed by the Xoxocotlan dam did
not contribute to the food supply of the considerably younger
Zapotec centre on nearby Monte Alban. Besides the famous religious and administrative structures, Monte Alban was the site of
an ingenious system to collect and store the precipitations as drinking water (O'Brien et al. 1980). At the time of its prime (from
about 250-650 AD) Monte Alban counted a population of 15 000
to 30 000 souls. The water supply system in part comprised covered channels of masonry, multiple settling tanks and small storage
dams in the gullies on the sides of the mountain. The latter were
also provided with check dams and terraces for cultivation.

Maya reservoirs

The reservoirs built by the Mayas in Yucatan were primarily
intended to satisfy the population's water requirements rather than
serving irrigation, hardly necessary in the tropical climate (Fig. 45).
The highest concentration of such reservoirs occurred in Tikal in
Guatemala, which flourished from about 600-900 AD. While
some reservoirs were excavated, others were impounded by dams
in small gullies, like the 50 000 m^3 – Palace reservoir (Fig. 47). Its
14 m high and 83 m long embankment had a cross-section which
appears quite modern. The volume of fill required amounted to
16 000 m^3.

Toltec irrigation dams

In parallel to the Zapotec and Maya civilizations the one of the
Toltecs flourished between 500 and 1100 AD approximately. Its
religious and administrative centre was Teotihuacan, 45 km northeast of Mexico-City (Fig. 45). The remains of three small irrigation
systems with reservoirs which were built during the final phase of
the Toltec civilization were located in the mountains west of
Teotihuacan (Armillas et al. 1956 and Millon 1957). The largest
Maravilla system included a 11 m high and 530 m long storage
embankment, a river training dam 200 m downstream, a diversion
weir some additional 600 m downstream and a 1.5 m wide canal
starting just above the weir. The holes preserved in the bedrock
show that the weir consisted of two 25 m long rows of posts,
which crossed the river diagonally. The posts in each row were
about 1.2 m apart and offset against those of the other row. Each
row supported a lattice work of branches and the 1.6 m wide
intermediate space was filled with earth and stones.

Anasazi storage schemes

Independently from these achievements in Mexico some tribes
in the southwest of the USA probably developed a considerable
skill in hydraulic engineering in order to survive in their arid
environment. Storage reservoirs were especially built by the Anasazi in southern Colorado and northern New Mexico, who have
become well known for their multistoried dwellings, the first 'sky-
scrapers' of the New World! They built an 8 km long canal along
the Chapin plateau in the Mesa Verde National Park around 1100

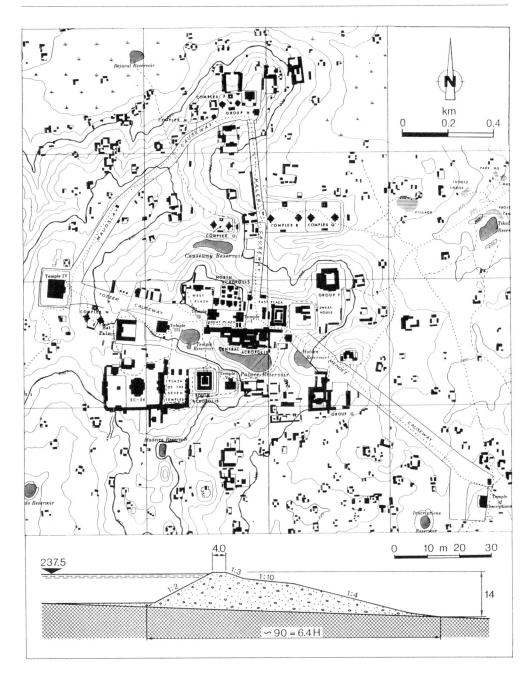

AD probably to bring additional water to a 1 m high and 9 m long dam across the head of the Fewkes Canyon at the southern end of the plateau (Fig. 45) (Rohn 1963). Near its head the canal had a short diversion to the circular Mummy reservoir, which measured 30 m in diameter and 3.6 m in depth. This reservoir was sur-

Figure 47. Map of the Tikal ruins in Guatemala and cross section of the Palace reservoir dam (map by R.F. Carr and J.E. Hazzard).

rounded by a stone masonry wall and the inlet from the canal had a sharp bend to have the sediments settle in it. From here they could be removed more easily than from the reservoir, which supplied water to nearby settlements. The chances are that it was older than the canal to Fewkes Canyon.

Pueblo Bonito and Gran Quivira systems

A similar canal, only 3 km long, brought water from a reservoir to Pueblo Bonito in the Chaco Canyon near the northwestern corner of New Mexico (Fig. 45) (Judd 1954). The scheme dates from the 12th-13th century AD. Still younger is the one of Gran Quivira on a plateau in the centre of New Mexico (Fig. 45) (Toulouse 1945). It comprised five reservoirs of 30 to 50 m diameter and 2 to 3 m depth, of which three were interconnected by short canals (Fig. 48). Due to the closeness of the reservoirs to the settlement and its fields only short supply canals were required.

A few centuries before the arrival of the Spanish conquerors in 1518 AD the Aztecs attained supremacy in Mexico. They started out from their capital Tenochtitlan in the centre of today's Mexico-City. In those days this was an island in the western part of Lake Texcoco, a lake which then covered most of the valley of Mexico-City (Figs. 45, 49). In the deeper eastern part its waters were nitrous and endangered the cultivation on artificial islands of mud and aquatic plants – the famous 'floating gardens', still to be seen in Xochimilco south of Mexico-City – which supplied the food to Tenochtitlan.

Aztec dams and dikes

As a remedy, the Aztec king Itzcoatl (1428-1440) had 9 and 6 km long dikes built from Tenochtitlan to the south and north respectively. These isolated the western part of Lake Texcoco and – together with a bifurcating western dike – provided road accesses to the city (Palerm 1973). Under king Montezuma I (1440-1469)

Figure 48. Plan of 5 reservoirs near Gran Quivira National Monument in New Mexico (after Toulouse 1945).

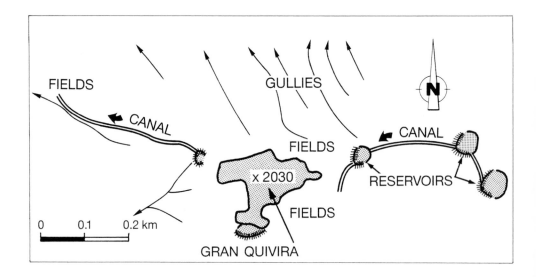

FIELDS

CANAL

GULLIES

N

CANAL

FIELDS

x 2030

RESERVOIRS

FIELDS

0   0.1   0.2 km

GRAN QUIVIRA

a 16 km long dam farther east was added. For its construction Nezahualcoyotl (1418-1472), king of Texcoco on the eastern shore of the lake, collaborated as hydraulic expert. The dam was therefore named after him. Like the Maravilla diversion dam, described above, it consisted of two rows of posts, between which large rocks were placed. During the Spanish conquest of Tenochtitlan in 1521 these dikes and dams were heavily fought over (Fig. 50).

Figure 49. Map of Lake Texcoco in Mexico with its ancient dams and dikes (after Palerm 1973).

1.9 SUMMARY

A listing of the oldest dam construction in each of the described regions of the world in chronological order yields the picture shown in Table 5. According to it, dam engineering was 'invented'

Figure 50. Fighting over a dam (underneath the right foot of the warrior in the upper left) from the Mexican Fernandez Leal Codex; a rare representation of warfare over dams (from Woodbury & Neely 1972).

Table 5. Chronological list of ancient dams.

Year completed	Country	Name of dam	Type	Function	Purpose
3000 BC	Jordan	Jawa	Gravity	Reservoir	Water sup.
2600 BC	Egypt	Kafara	Embank.	Reservoir	Flood contr.
2500 BC	Baluchistan	Gabarbands	Gravity	Reservoir	Conserv.
1500 BC	Yemen	Marib	Embank.	Diversion	Irrigation
1260 BC	Greece	Kofini	Embank.	Diversion	Flood contr.
~1250 BC	Turkey	Karakuyu	Embank.	Reservoir	Water sup.
950 BC	Israel	Shiloah	?	Reservoir	Water sup.
703 BC	Iraq	Kisiri	Gravity	Diversion	Irrigation
700 BC	Mexico	Purron	Embank.	Reservoir	Irrigation
581 BC	China	Anfengtang	Embank.	Reservoir	Irrigation
370 BC	Sri Lanka	Panda	Embank.	Reservoir	Irrigation
275 BC	Sudan	Musawwarat	Embank.	Reservoir	Water sup.

in the first half of the third millennium BC in southwestern Asia and northeastern Africa, primarily for the purpose of water supply, flood control as well as soil and water conservation. Irrigation became a purpose only one millennium later. In the last millennium BC dams were build all over the world, from China to Mexico. Except for the abovementioned, possible connection between Urartian (Turkey) and Assyrian (Iraq) hydraulic engineering, the construction of dams in the various regions of the world developed simultaneously or, as Needham wrote: 'Where the technological elements of water, earth and wood are concerned, people in very different parts of the world (may well have) developed their own traditions fairly independently – for metal I would not like to say as much.' (Needham 1971).

As varied as their origins, were the structural characteristics of the ancient dams, for which no regional preferences are discernible. They had one aspect in common however: to resist the water pressure alone the weight of the construction material was used, but not its strength (other than that required to keep it from crushing under its own weight). According to modern classification they were all either embankment or gravity dams. The arch and buttress types were to be introduced only by the Romans (see below). While all the embankments, except Sadd-el Kafara, were homogeneous, i.e. had no impermeable core, the following subtypes can be distinguished among the gravity dams:

– solid, mostly rectangular masonry walls up to 19 m high (Asid in Yemen);

– two masonry walls with an intermediate earth core up to 15 m high (Keşiş Gölü in Turkey); and

– a masonry wall – with or without core – backed up by a downstream fill up to 7 m high (Mala'a in Egypt); this type can also be regarded as an embankment with an upstream water retaining wall.

It is difficult to say which one of the 'pure' ancient embankments was the highest because it is unknown when the Marib dam in Yemen reached its final height of 20 m. Datable are the 19 m attained by the Wahalkada dam in Sri Lanka in the first century AD and the record for the next 1200 years established in 460 AD in the

Dams all over the ancient world

Independent developments

Only structures resisting by weight

Various types of gravity dams

Record height embankments

Table 6. Ancient record reservoir capacities.

Year completed	Country	Name of dam	Capacity (million m^3)
13th c. BC	Greece	Boedria	24
~700 BC	Turkey	Keşiş Gölü	100
581 BC	China	Afengtang	100
3rd c. BC	Egypt	Mala'a	275

Table 7. Record periods of operation of ancient dams (over 2000 years).

Country	Name of dam	Period of operation	
		Years	End
Egypt	Mala'a[1]	3600	~1900 AD
Greece	Kofini	3300	Still operating
Turkey	Keşiş Gölü	2600	1891 AD
China	Anfengtang	2600	Still operating
Sri Lanka	Bassawak	2300	Operating again
Sri Lanka	Tissa (Anurad.)	2300	Operating again
China	Tianping	2200	Still operating
Sri Lanka	Pavat	2200	Operating again
Sri Lanka	Vavuni	2200	Operating again
Yemen	Marib[1]	2100	630 AD
Sri Lanka	Nuwara	2100	Operating again
Israel	Solomon Pools	2000	Still operating

[1] Incl. preceeding structures.

Considerable storage capacities

Long periods of operation

same country by the second heightening of the Paskanda Ulpotha dam to a total of 34 m.

Most of the ancient dams impounded storage reservoirs, some of which had astonishing capacities (Table 6). Not less remarkable are the long periods of time, during which some of the dams were more or less continuously in operation (Table 7). Those which fell into ruin, decayed rather rarely due to technical deficiencies or extraordinary natural events, like floods or earthquakes. Most fell prey to socioeconomic or political changes, after which maintenance (e.g. removal of sediments) and/or repair were neglected.

CHAPTER 2

The Roman empire

2.1 GENERAL AND ITALY

Possibly inspired by their northern Etruscan neighbours who were proficient hydraulic engineers (Bonnin 1973), the Romans undertook major hydraulic works quite early. Thus, at the beginning of the fifth century BC they drained the centre of their city, the Forum, with the Cloaca Maxima sewer. At the same time they dewatered the Ariccia valley by a 607 m long tunnel 30 km southeast of Rome and, in 396 BC, they lowered the level of the nearby Albano lake through a 1200 m long tunnel to gain cultivable land (Grewe 1981). From 312 BC until 52 AD the Romans also built nine aqueducts with a total length of 423 km to supply their city with water. From 20 BC (at Nîmes/France) onwards they exported this technique also to their provinces.

Early Roman hydraulic works...

It is therefore astonishing that the Romans took up dam engineering at a relatively late date. When they annexed Greece around 150 BC the Mycenaean dams (see above) were probably already all silted-up and the few dams from the fifth/fourth centuries BC were too insignificant. In the middle of the first century BC the Romans installed themselves in the Near East and thus became acquainted with Israelite, Nabatean and Ptolemaic dams. Moreover an unsuccessful campaign in 25/24 BC took them as far as Marib in Yemen and its great dam which was devastated by them (see above).

...but late dams

When the Romans began to build dams they possessed fully developed construction techniques, which were based on manual labour with simple tools, such as levers, picks and shovels, like in the other ancient civilizations. The labour force consisted of slaves of private contractors or, in public works such as hydraulic schemes, very often of idle troops. For transportation of the construction materials man or animal drawn carts or stretchers carried by two men were used, but not as yet the wheelbarrow which is a medieval import from China (Matthies 1991). Mechanical devices

Mature construction techniques

Construction machinery

were also employed for pulling and lifting by rope, such as a single pulley or several of them combined to a tackle from about 330 BC (Kottmann 1979). For lifting they were hung from tiedback inclined poles or the horizontal arms on top of the vertical standards of cranes. The rope was mostly wound up on a windlass turned by cross-bars (not a crank which was another medieval invention) or, from about 230 BC by a treadmill wheel (Fig. 51).

Several instruments were used by the Romans to survey and stake out their structures (Grewe 1985). They had rulers, squares and compasses for drafting, abacuses for calculations as well as rulers, plumblines, spirit levels and measuring rods for fieldwork.

Surveying methods

Right angles were staked out with a horizontal cross, from the four ends of which hung plumblines. Similarly, plumblines and marks on the supporting structure served to level out instruments that were about 6 m long and which were moreover provided with a water-level cut into their surface. However this cumbrous instrument was used only for the basic surveys, while intermediate points were levelled by sighting over poles with horizontal arms on their tops.

Like their predecessors, the Romans used earth, stones and wood as construction materials. In addition, for the first time they employed concrete, a mixture of sand and gravel, burnt lime, water and volcanic ash or ground brick on a large scale (Lamprecht 1984, Malinowski & Garfinkel 1991). The last mentioned ingredients permitted the concrete or mortar to harden under water and to resist leaching out by it. Since concrete could not only be poured between masonry walls but also into temporary wooden formwork or shuttering it allowed the building of structures of any shape, like for example, the impressive vaults, so dear to the Ro-

Construction materials, especially concrete

Figure 51. Reconstruction of a Roman crane, the windlass of which is driven by a treadmill wheel, in the Cimiez baths above Nice in France (photo by the author).

mans. But they also used concrete extensively in their dams, since it was cheaper than masonry wherever the width of the wall exceeded a certain measure.

Incidentally all Roman dams in principle were simple walls of constant width which were founded on rock and, when of moderate height, overflown by floods (Fig. 52). Rarely one or both faces of a dam were inclined, because it was only in the 19th century when engineers realized, that the optimum cross section would be a triangle corresponding to the increase of the water pressure in the reservoir from top to bottom, a fact already established by Archimedes of Syracuse (287-212 BC) in Sicily. It was however the same scientist who considered all art which satisfied a need, as a matter of the lowly artisans (Plutarch). This attitude was typical for the ancient separation between science and technology, which was to last until well after the Middle Ages.

Whenever the Roman engineers, often erroneously, judged the stability of a dam wall to be insufficient, they backed it up by

Principle of Roman
dams

Use of buttresses

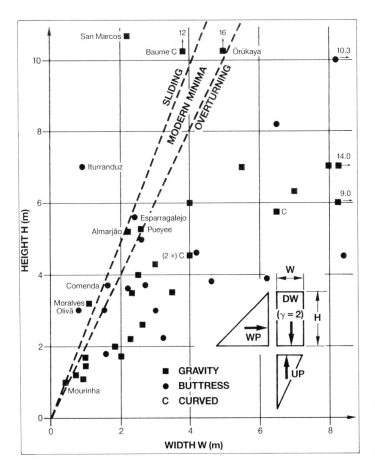

Figure 52. Width of Roman gravity and buttress dam walls in function of their height; DW = dead weight, UP = uplift, WP = water pressure; critical dams identified by name.

Rare arch dams

Nero's dam at Subiaco

irregularly spaced buttresses (Fig. 52) or, especially at higher dams, by an embankment on the downstream side. Very rarely did they curve their dams in plan to gain additional stability through moderate horizontal arch action.

This would have been quite helpful for the 40 m high and 13.5 m wide dam built for a pleasure lake near his villa at Subiaco, 50 km east of Rome, under emperor Nero (54-68 AD) (Fig. 53) (Smith 1970). Under Trajan (98-117) the intake for the Anio Novus aqueduct to Rome built between 38 and 52 was moved to Nero's lake. The dam was one of the earliest Roman dams and it was also to remain the highest they ever built. Moreover, it and two smaller structures up- and downstream are the only Roman dams so far known in Italy. Its crest was straight but only some 80 m long. Due to wedge action in the narrow canyon it resisted despite its

Figure 53. Ruins of two spill-way piers on the right abutment of emperor Nero's Subiaco dam (photo by the author).

Figure 54. Painting in the Sacro Speco monastery showing the Subiaco dam before its failure in 1305 (photo by A. Vogel, Vienna).

underdesigned width of only 34% of the height until 1305 AD, when monks from the nearby Sacro Speco monastery removed stones from the dam crest. In the monastery there is a painting of 1428 AD showing the dam before its failure (Fig. 54).

2.2 THE WESTERN PROVINCES

The first provinces which the Romans acquired outside of Italy were the Carthaginian territories along the eastern and southern coasts of the Iberian peninsula around 200 BC. Although the Romans started to build dams there only towards the end of the first century AD we shall begin by describing them because they are the most numerous and best researched (Table 8) (Cabalero & Sánchez-Palencia 1982, Díaz-Marta 1992, Fernández et al. 1984, Quintela et al. 1986). The Roman dams – about 80 of them – identified so far in the region can be divided into seventeen groups all concentrated along the principal Roman roads (Fig. 55). About half of them were used for irrigation, two fifths for water supply and the rest for power supply in mines, to either operate pumps or lifting devices. However, two fifths of the dams simultaneously served two or more purposes, like irrigation and water supply for larger farms. Half of the dams simply were gravity walls, while one third were backed-up by downstream buttresses and the rest had downstream embankments. The reservoir content of the low dams went from a few thousand m^3 to some 100 000 m^3 whereas it attained millions of m^3 only at the high dams (Fig. 56).

Many Roman dams on the Iberian peninsula

Purpose and type of the dams

2.2.1 *Large embankments*

On the Iberian peninsula, all these high Roman dams were embankments. One of the oldest was probably the Alcantarilla dam built some 20 km south of Toledo in central Spain as head pondage for the 50 km long aqueduct to that city. Just before entering Toledo it crossed the Tajo river on one of the mightiest inverted siphon bridges the Romans ever built (Smith 1976). The reservoir of 3.5 million m^3 capacity was filled by the runoff from a 50 km^2 area of direct catchment and that of 42 km^2 of a creek to the west, which was brought to the reservoir by a canal (Fernández 1961, Celestino 1974).

Alcantarilla embankment

The dam followed the remains of a natural barrier across the valley with two changes in the alignment and was about 500 m long. It consisted of an upstream wall of up to about 17 m in height and a downstream embankment. The wall was vertical on the upstream and slightly inclined on the downstream side, so that its width increased from 3.2 m at the crest to some 6 m at its base. It

Structural details

Table 8. Roman dams in southwestern Europe (of which the dimensions are known).

Nearest city (country)	Name of dam	Type	Height (m)	Length (m)	Base width (m)	Purpose
Beja (Portugal)	Alamo	Buttress	3.0	50	3.0^2	Irr./Water s.
	Castelinho	Embankment	0.8	56	11.0	Irr./Water s.
	Fonte Coberta	Gravity	2.6	75	2.6	Irr./Water s.
	Santa Rita	Buttress	2.2	50	3.2^2	Irrigation
	Tesnado	Gravity	1.2	220	0.7	Irr./Water s.
Evora North (Portugal)	Almarjão	Gravity	5.2	55	2.2	Irr./Water s.
	Carrão	Gravity	1.7	117	1.0	Irr./Water s.
	Gavião[1]	Buttress	10.0	78	10.3	Irr./Water s.
	Moralves	Gravity	3.2	161	1.1	Irrigation
	Mourinha	Gravity	1.0	100	0.4	Irrigation
	Muro	Buttress	4.6	174	4.2^2	Irr./Water s.
	Olivã	Buttress	3.0	45	0.8^2	Irrigation
	Tapada	Embankment	1.6	76	13.2	Irrigation
Evora South (Portugal)	Baleizão	Gravity	1.1	120	0.9	Irrigation
	Cuba	Buttress	1.8	81	1.6^2	Irrigation
	Monte Novo[1]	Arch	5.7	52	6.5	Power
	Mouros	Buttress	3.0	130	1.5^2	Irrigation
	Pisões	Gravity	4.3	58	3.0	Irr./Water s.
	Prega	Buttress	3.9	62	6.2^2	Irr./Water s.
Granada (Spain)	Barcinas	Gravity	4.5	40	4.0	Irrigation
	Deifontes	Gravity	4.5	15	4.0	Irrigation
Lisbon (Portugal)	Comenda	Buttress	3.7	13	1.6^2	Irr./Water s.
	Olisipo	Buttress	8.2	52	6.5^2	Water supply
Merida (Spain)	Araya	Buttress	3.7	139	2.7^2	Irr./Water s.
	Cornalbo	Embankment	24.0	220	~90	Water supply
	Esparragalejo	Mult. Arch	5.6	320	2.4^2	Irrigation
	Hinojal	Buttress	1.5	250	?	Irr./Water s.
	Proserpina	Embankment	21.6	426	~70	Water supply
	Santa Maria	Buttress	3.6	98	2.2^2	Irrigation
	Tomas	Embankment	3.0	95	?	Irr./Water s.
Pamplona (Spain)	Iturranduz	Buttress	7.0	102	0.9^2	Water supply
St. Rémy (France)	Baume	Arch	12.0	18	3.9	Water supply
Toledo (Spain)	Alcantarilla	Embankment	17.0	557	~60	Water supply
	Consuegra	Embankment	5.8	700	?	Water supply
	Melque I	Gravity	5.0	60	5.8	Power/Water s.
	Melque II	Gravity	2.2	57	2.3	Power/Water s.
	Melque III	Gravity	1.7	11	2.0	Power/Water s.
	Melque IV	Gravity	9.0	60	14.0	Power/Water s.
	Melque V	Gravity	6.0	40	4.0	Power/Water s.
	Moracanta	Gravity	2.0	40	1.8	Irr./Water s.

Table 8 (Continued).

Nearest city (country)	Name of dam	Type	Height (m)	Length (m)	Base width (m)	Purpose
	Paeron I	Embankment	2.4	81	?	Irr./Water s.
	Ponton Grande	Buttress[3]	4.5	58	8.4^2	?
	Ponton Chico	Buttress[3]	3.8	25	4.6^2	?
	Valhermoso	Embankment	3.0	98	?	Irr./Water s.
Zaragoza (Spain)	Muel	Gravity	13.0	?	?	Irr./Water s.
	Pueyee[1]	Gravity	5.3	47	2.6	Irrigation
	San Marcos[1]	Gravity	10.7	33	2.2	Irr./Power

[1]Roman origin uncertain. [2]Width of water retaining wall only. [3]With earth core.

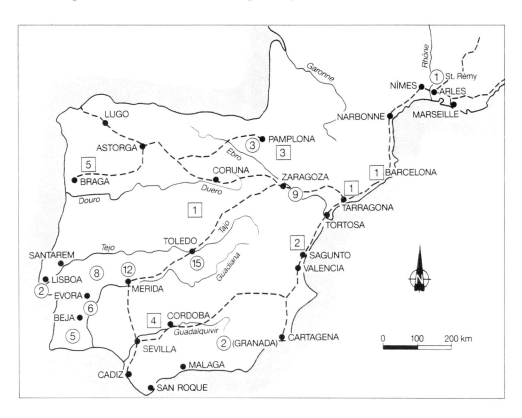

was made up of two 1.0 m thick rubble masonry walls and an intermediate concrete core. Moreover, the upstream face was lined with 0.5 m thick ashlar masonry. The embankment was 14.0 m wide at the crest and had a downstream slope of about 1:3. Near the highest part of the wall one, or possibly two, access shafts to outlets through the dam's base were built against its downstream face. They had 1.0 m thick walls of rubble masonry and a clear opening of 4.0 by 4.0 m, which contained a spiral staircase.

Figure 55. Map of the Iberian peninsula and southern France in Roman times with location of the main cities (present names), the principal roads (broken lines) and the groups of dams (circles with the number of dams in each group; squares = only vague information available).

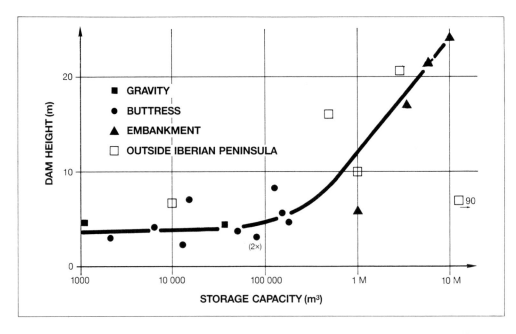

Figure 56. Storage capacity of Roman dams in function of their height.

Failure of Alcantarilla dam

Improved Proserpina embankment

Outlet works

Spillways were provided by overflows at both ends of the dam. Due to their insufficiency the centre of the dam was breached by overtopping and its remainder was pushed upstream by the embankment after the sudden emptying of the reservoir (Celestino 1974). Others believe that it happened after a more or less normal emptying due to soaking of the embankment by rain or seepage through the water retaining wall (Fernández 1961). In any case it is a striking fact that all its debris lies on the upstream side (Fig. 57).

To prevent the same thing from happening at the very similar Proserpina dam, 5 km northwest of Merida in southwestern Spain, the highest part of its upstream face was provided with nine 0.7 m thick buttresses spaced about 20 m apart (Fig. 58) (Arenilla et al. 1992, Castro 1933, Celestino 1980, Fernández 1961). Moreover, not only the downstream face of the water retaining wall was inclined but even more so the upstream one. Its width increased from 1.6 m at the crest to 5.7 m about 15 m below it. As discovered on the occasion of a recent removal of the sediments from the reservoir, the lowest 6 m of the wall had a constant width of 5.7 m and constituted a 150 m long first stage of the dam. The internal structure of the wall as well as the configuration of the supporting embankment were the same as at Alcantarilla.

Of almost identical construction and size were the two access shafts to the outlets through the Proserpina dam. The deeper one reached down to two lead pipes of 0.22 m diameter embedded in the first stage wall, which were later replaced by new openings higher up. From these the water was led in an 8 km long open canal

Figure 57. Ruins of the Alcantarilla dam south of Toledo in Spain; to the left of the water retaining wall is one of the outlet towers (photo by the author).

Figure 58. Crest and upstream face of Proserpina dam northwest of Merida in Spain with one of the upstream buttresses; the decoration on the latter's top dates from the 17th century (photo by the author).

to the remarkable Los Milagros (The Miracles) aqueduct bridge and into Merida, most probably for industrial use. The direct catchment area of 7 km^2 upstream of the dam was insufficient to fill the reservoir of 6 million m^3 capacity. Therefore additional water was collected from 13 km^2 of adjoining watersheds by a 3.5 km long canal in this case as well. Thanks to careful maintenance and several major repairs, the Proserpina dam is still in operation, and this roughly 1900 years after its construction! It now serves mainly for local irrigation and recreation.

Also still in use is the Cornalbo embankment dam 12 km northeast of Merida, to where it originally supplied drinking water through a 20 km long underground aqueduct (Fernández 1961, Celestino 1980). With a height of 24 m it is also the highest dam the Romans built outside of Italy (see above) and their largest with almost 70 000 m^3 of volume. Its reservoir capacity amounted to some 10 million m^3.

Cornalbo embankment

The most remarkable aspect at Cornalbo however is the precautions taken against upstream sliding by tilting the water retaining wall downstream, a measure which necessitated its support by a system of transversal and longitudinal masonry walls (Fig. 59). The cells thus formed were filled with stones and clay as in the case of the Precolumbian Purron dam in Mexico (see above). Then they were covered with concrete and an ashlar revetment.

Sloping upstream face

Because of the inclination of the water retaining wall, the access shaft to the 0.5 m wide and 1.7 m high outlet conduit through the dam's base, became an intake tower standing inside the reservoir, a rare solution in old dams but quite frequent in modern ones. A double arch bridge that has since disappeared, led from the dam crest to the top of the shaft. Again there was a spiral staircase in its clear interior of 4.5 by 4.5 m. Its walls of ashlar masonry were only 0.5 m thick in their upper 10 m.

Figure 59. Cross section through the outlet of Cornalbo dam northeast of Merida in Spain (after Fernández et al. 1984).

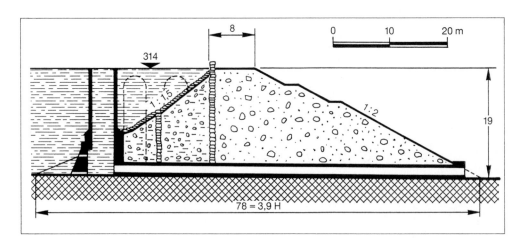

2.2.2 *Novel buttress dams*

As mentioned earlier the Romans introduced downstream buttresses instead of embankments in their lower and shorter dams to
ensure or increase (very often unnecessarily) the stability of the
water retaining wall (Fig. 52). At the 5.8 m high and 700 m long
Consuegra dam, 57 km southeast of Toledo in central Spain, both
methods of stabilization, in this case imperative, were used simultaneously (García-Diego 1980). By comparison the buttresses at
the Ituranduz dam 28 km southwest of Pamplona in northeastern
Spain were insufficient. Excavations in 1980 showed that the water
retaining wall of 102 m in length and up to 7 m in height, only
measured 0.9 m in width (Mezquiriz 1984). Therefore, the 2.2 m
thick and 2.1 m wide buttresses spaced 5.1 m apart (centre to
centre) along the highest section of the dam were not enough to
safely prevent it from sliding on the foundation rock. The central
part of the dam failed at an unknown date, probably in connection
with overtopping by a flood, all the more as it had no separate
spillway.

    By contrast, the 8.2 m high and 52 m long Olisipo buttress dam
10 km northwest of Lisbon in Portugal was rather overdesigned
(Fig. 60) (Almeida 1969). Its 125 000 m^3 reservoir constituted the
head pondage for an aqueduct to Lisbon. The 6.5 m wide, rectangular water retaining wall was of course stable by itself. It consisted of concrete poured between two walls of crudely dressed
ashlar masonry. The 5.1 m wide and approximately 2.2 m thick
buttresses were built in the same manner and were spaced some
4.9 m apart.

    Doubtlessly one of the most remarkable Roman buttress dams
is the one near the village of Esparragalejo, 9 km northwest of
Merida in southwestern Spain. It was rebuilt in 1959 from its
substantial remains (Fig. 61) (García-Diego 1971-1972). The 5.6
m high and 320 m long structure was provided in its central part
with twelve buttresses of an average width of 1.2 m, 3.2 m thickness and 8.6 m spacing. However, the novel feature was that the
downstream face of the at least some 2 m wide water retaining wall
was no longer straight but curved between the buttresses. Thus, the
first multiple arch dam was born!

2.2.3 *First arch dams*

It is a rather astonishing fact that the Romans only sparcely applied
the arch in dam constructions, a design otherwise so masterfully
employed in their buildings and bridges. Its cautious application at
the Esparragalejo multiple arch dam has just been mentioned. In
addition, the Romans curved four of their 25 gravity dams so far
identified on the Iberian peninsula in plan towards upstream.

Buttresses in the
Consuegra embankment

Insufficient buttresses at
Ituranduz

Sturdy Olisipo dam

Esparragalejo multiple
arch dam

Rare arch dams

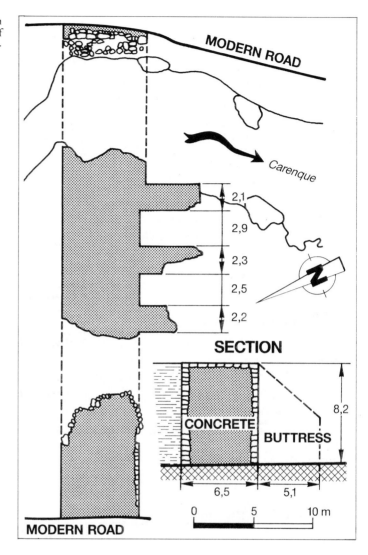

Figure 60. Plan and cross section of the Olisipo dam northwest of Lisbon in Portugal (after Almeida 1969).

Barcinas curved weir

One of these was the Barcinas diversion weir on the Cubillas river, 20 km north of Granada in southeastern Spain (Fernández et al. 1984). It had a rectangular cross section of up to 4.5 m height and a crest length of 40 m. The left half thereof which crossed the river, was strongly curved with a radius of 12 m and a central angle of about 100°. On the left side the arch abutted against the rocky river bank, while on the right side it continued into a straight wing wall. It seems however that in this case the arching was not adopted to take advantage of its anyhow modest stabilizing effect. It rather served the lengthening of the dam crest for easier passage of floods.

A true arch dam was built by the Romans in the Vallon de Baume 4 km south of Saint-Rémy de Provence in southeastern

Figure 61. View along the crest of the Esparragalejo multiple-arch dam northwest of Merida in Spain, as rebuilt in 1959 (photo by the author).

France for the water supply of the city (Benoit 1935). The site was almost entirely covered by a modern dam in 1891, but fortunately a rather precise plan of the Roman foundation excavations from 1765 still exists (Fig. 62). Accordingly the dam was some 12 m high and 18 m long and curved with a radius of about 14 m and a central angle of 73°. It consisted of two masonry walls, 1.3 m wide on the upstream and 1.0 m on the downstream side, with an intermediate earth core of 1.6 m in width. Therefore, the downstream arch had to carry the whole water load which, even considering the relatively small central angle, stressed it at mid-height to only 1.2 MPa near the downstream abutments.

Another true arch dam was found at Monte Novo, around 15

Vallon de Baume true
arch dam

Figure 62. Plan of the Baume arch dam near Saint-Rémy de Provence in southeastern France plotted into the 18th century survey of its foundation excavations (after Benoit 1935).

Possibly Roman arch dam

km east of Evora in southeastern Portugal (Quintela et al. 1986). Only recently it was unfortunately submerged by a modern reservoir. Further investigations were thus foiled, namely the determination of whether it actually was of Roman origin. On the other hand who else but the Romans could have built such a sophisticated structure up to very recent times (Fig. 63)? It was 5.7 m high and 52 m long, including the wing walls at both ends. The middle part was curved with a radius of 19 m and a central angle of some 90°. The dam was built of blocks of shist laid horizontally in lime mortar.

The refinement of the Monte Novo dam consists of the fact that the curved central part at both ends passed into the wing walls

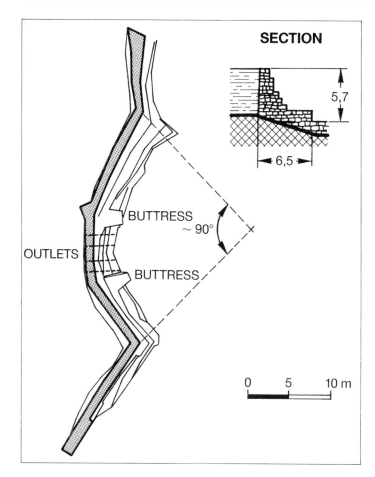

SECTION

5,7

6,5

BUTTRESS

~ 90°

OUTLETS

BUTTRESS

0   5   10 m

Figure 63. Plan and section of the Monte Novo arch-gravity dam in southeastern Portugal (after Quintela et al. 1986).

without any abutment blocks to absorb the horizontal arch thrusts. A similar design was adopted again as late as in 1300 for the Kebar arch dam in central Iran (see below) and in the upper parts of several modern arch dams. It works, because the total stress trajectories in arch dams dip towards the abutments. However, the builders of the Monte Novo dam appear not to have entirely relied on their bold design because they added two buttresses to its downstream face. They also provided two outlets of 1.2 and 1.4 m width respectively and 1.0 m height through the dam's base. The triangular platform projecting upstream from the crest might have been intended to check the good functioning of the outlet gates. The released water is said to have driven mills farther downstream.

Sophisticated design

2.3 NORTH AFRICA

After their final victory over the Carthaginians in 146 BC, the Romans installed themselves in Algeria, Libya, Morocco and

Many Roman dams also in North Africa

Tunisia as well. In these African provinces the Romans took up dam building probably somewhat later than on the Iberian peninsula, i.e. mainly in the second century AD. The structures of which the dimensions are known, are summarized in Table 9 (Gauckler 1897-1902, Gsell 1902, Vita-Finzi & Brogan 1965). Aside from the twelve dams listed in the table, remains of some 120 additional structures have been reported (Fig. 64). The vast majority of them was used for irrigation or water and soil conservation, in the manner the Nabateans had perfected it in southern Israel and Jordan and which the Romans learnt to know in the first century AD (see above).

Mostly masonry walls

Most North African dams were solid masonry walls, often with a concrete core or even built completely of concrete. The downstream face on some was stepped to average inclinations between 20 and over 100%. Since the base widths of these dams amounted to at least 70% of their height and often exceeded this figure considerably, their stability was ensured. The smallest inclination and relative base width was evident at the 10 m high Derb dam near Kasserine, 220 km southwest of Tunis (Saladin 1886). It was some 130 m long and curved in plan. Its width measured 7.0 m at

Large Derb dam

the base and 4.9 m at the crest. The dam was built of concrete lined with ashlar masonry. In modern times the French reconstructed the partly ruined structure during their occupation of Tunisia (Fig. 65).

Table 9. Roman dams in North Africa (of which the dimensions are known).

Nearest city (country)	Name of dam	Type	Height (m)	Length (m)	Base width (m)	Purpose
Barika (Algeria)	Barika	Gravity[1]	3.2	50	7.0	Irr./Water s.
	Roumila	Gravity[1]	2.5	25	4.0	Irrigation
	Sakhri	Gravity	1.1	15	2.5	Irrigation
Collo (A)	Halla	Gravity	5.0	?	?	Irrigation
El Attaf (A)	Taria	Gravity[1]	4.0	?	3.5	Irrigation
Ighil-Izane (A)	Djidiouïa	Gravity[1]	5.0	?	5.0	Water supply
Kasserine (Tunisia)	Derb	Gravity[1,2]	10.0	~130	7.0	Irrigation
Mila (A)	Mila	Gravity	10.0	70	?	Irrigation
Sidi Kada (A)	Kerroucha	Gravity	6.0	?	9.0	Irrigation
Tarabulus (Libya)	Megenin I	Buttress	6.3	91	5.4	Irrigation
	Megenin II	Gravity[1]	5.8	257	4.3	Irrigation
	Megenin III	Buttress	1.5	43	1.0	Irrigation

[1]Downstream face inclined. [2]Curved in plan.

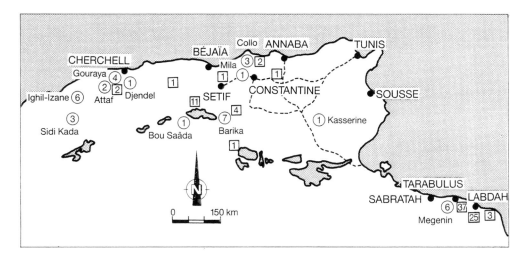

Figure 64. Map of western North Africa in Roman times with locations of the main cities (present names), the principal roads (broken lines) and the groups of dams (circles with number of dams in each group; squares = only vague information available).

Figure 65. Downstream view of the reconstructed Derb dam in central Tunisia and its original Roman section (photo by the author, section after Saladin 1886).

Some important masonry dams were built by the Romans for irrigation in the Megenin valley some 40 km south of Tarabulus (formerly Tripoli) in Libya (Vita-Finzi & Brogan 1965). In total there were six dams which partly succeeded each other. The first was an up to 6.3 m high and 91 m long, sturdy wall with about six buttresses along the downstream face. These were actually unnecessary as in many of the Roman dams on the Iberian peninsula. An interesting perfection of the technique was a 0.02 m thick plastering of the upstream face with lime mortar. After the river changed its course, a similar but much longer second dam was built upstream of the first one. This time, no buttresses were provided and the downstream face was stepped to an average inclination of about

Megenin dams

Figure 66. Upstream view of a
curved, earth backed retaining
wall on the Kl'am torrent
southeast of Labdah in Libya
(from Vita-Finzi 1961).

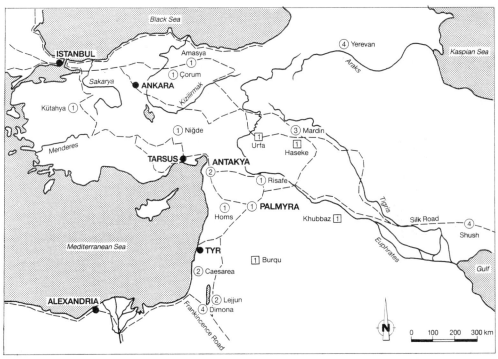

Figure 67. Map of the Near and
Middle East in Roman times
with locations of the main cities
(present names), the principal
roads (broken lines) and the
groups of dams (circles with
number of dams in each group;
squares = only vague information available).

30%. Moreover the new dam got a much ampler spillway than the
old one.

Some 65 Roman dams were identified around Labdah, 115 km
east of Tarabulus, home town of the emperor L. Septimius Severus
(193-211 AD) (Fig. 64) (Vita-Finzi 1961). A large buttress dam
diverted the muddy flood waters of the Labdah into a canal by-
passing the city and its harbour. An aqueduct about 20 km long was
built for the water supply from springs in the Kl'am valley to the
east where a collecting reservoir was constructed. The latter was

protected from floods and sediments by the many soil and water conservation dams erected further upstream across the torrent and its tributaries. Among them there were also some earth backed retaining walls of the kind described earlier on the Iberian peninsula (see above) (Fig. 66).

Labdah river diversion

2.4 THE EASTERN PROVINCES

Although the Romans learnt to know dam engineering for the first time in their eastern provinces, they applied this knowledge also there only from about the end of the first century AD onwards. Like on the Iberian peninsula most dams in the east were located near the principal Roman roads connecting in the south with the ancient 'Frankincense Road' through Arabia (see above) and, in the east, with the more recent 'Silk Road' to China (Fig. 67). The dams of which the dimensions are known, are listed in Table 10 (Brossé 1923, Garbrecht 1991a, b, Garbrecht & Vogel 1991, Peleg 1986, 1991, Schlumberger 1939, Stark 1957-1958, Vries 1987)

Location of Near Eastern dams

Table 10. Roman dams in the Near and Middle East (of which the dimensions are known).

Nearest city (country)	Name of dam	Type	Height (m)	Length (m)	Base width (m)	Purpose
Amasya (Turkey)	Löştüğün[1]	Embankment	12.0	70+30	68	Irrigation
Antakya (T)	Çevlik	Gravity	16.0	49	5.0	River divers.
Caesarea (Israel)	Maagan North	Gravity	3.5	900	2.5	Water supply
	Maagan West	Gravity	7.0	193	5.5	Water supply
Çorum (T)	Örükaya	Gravity[2]	16.0	40	5.0	Irr./Flood c.
Dimona (I)	Churvat Zafir	Gravity	4.5	15	2.0	Water supply
	Maaleh Zafir	Gravity	2.9	13	1.0	Water supply
	Nachal Zafit	Gravity	3.5	22	3.5	Water supply
	Nachal Zin	Gravity	2.5	15	2.0	Water supply
Homs (Syria)	Qadas	Gravity	7.0	2000	14.0	Irrigation
Kütahya (T)	Çavdarhisar	Gravity[2]	7.0	80	8.0	Flood control
Lejjun (Jordan)	Upper	Gravity[2]	6.3	80	7.0	Water supply
	Lower	Gravity	3.5	100	3.5	Water supply
Mardin (T)	Dara East	Arch?	5.0	190	4.0	Flood control
	Dara West	Gravity[2]	5.0	66	?	Water supply
Niğde (T)	Böget	Gravity[2]	4.0	300	2.5	Water supply
Palmyra (S)	Harbaqa	Gravity	20.5	365	18.0	Irr./Waters s.
Risāfe (S)	Sêlé	Embankment	3.0	480	?	Water supply

[1]Roman origin uncertain. [2]With earth core.

2.4.1 *More gravity dams*

Çevlik river diversion

Like in North Africa the simple gravity wall with a mostly rectangular cross section was the preferred dam type in the eastern provinces of the Roman empire as well. The oldest is a dam not unlike the one at Labdah in Libya (see above) and which diverted muddy flood waters into a 130 m long tunnel and a 700 m long canal in rock around Çevlik, the harbour of the important city of Antakya in southeastern Turkey (Garbrecht 1991a). Due to two inscriptions along the canal the completion of the scheme can be dated to about 80 AD. The dam was some 60 m long and consisted of two 0.5 m wide walls of crudely dressed ashlar masonry and a 4.0 m wide concrete core. Given its height of 16 m (whereof 12 m are silted up) one wonders why the dam is still standing.

Precarious stabilities

The Örükaya dam, an analogous construction near Çorum 190 km northeast of Ankara in central Turkey is cause for even more astonishment (Stark 1957-1958). It had a 3.6 m wide core of earth (!) between two 0.7 m wide ashlar masonry walls, of which the downstream one has disappeared in the upper two thirds. Although the reservoir is also totally silted up, the remaining parts of the dam are still intact. This is possibly due to its being wedged into a gorge 40 m wide at crest level and only 12 m at the dam's base. At the base a vaulted outlet with some 3 m^2 of clearance could be closed by a wooden gate.

Çavdarhisar flood retention dam

Two similar Roman dams of lower but much sturdier design were found near Kütahya and Niğde 290 km west and 270 km southeast of Ankara respectively (Stark 1957-1958). The first one was provided with a large bottom outlet with a clear section on 11 m^2 which could not be closed by any gate available at the time (Fig. 68). It follows that this Çavdarhisar dam was most probably

Figure 68. Upstream view of the Çavdarhisar flood retention dam near Kütahya in western Turkey (photo H. Fahlbusch, Ratzeburg/D).

not intended to store water but only to temporarily retain floods and to break their impetus. The 300 m long Böget dam for the water supply to Misli near Niğde skillfully followed a ridge in the middle of the valley, although this implied an unusual downstream curvature of the structure. Its stability was increased by flat earthfills up to half the height on both faces.

Böget dam

Two lower dams were built by the Romans in the Lejjun ravine for the water supply to the nearby frontier fortress about 80 km south of Amman in Jordan (Fig. 67) (Vries 1987). The upper one was unusual in that its two outer walls were not parallel but diverging towards the abutments. Thus, the width of the earth core increased from a minimum of 2 m to 10 m at the left and even 30 m (!) at the right end. The walls consisted of ashlar masonry and measured 2.2 and 2.8 m of width respectively. The lower Lejjun dam was a simple masonry wall with a rectangular cross section. Its height of 3.5 m was equal to the width and the length amounted to roughly 100 m. Four similar water supply dams for Roman fortresses were found southwest of Lejjun near Dimona, some 80 km south of Jerusalem (Peleg 1991). Unfortunately no details are available on the dams near the fortresses of Burqu and Quasr al-Khubbaz between Ammann and the River Euphrates (Stein 1940).

Dams for frontier fortresses

The rich city of Palmyra, 215 km northeast of Damascus in Syria also formed part of the Roman frontier outposts (Fig. 67). A large dam was built 80 km southwest of it. Its purpose was to grow agricultural products on irrigated land rather than the supply of water to the city (Fig. 69) (Schlumberger 1939). The impressive structure was 20.5 m high and 365 m long. Its base width measured 18 m or 88% of the height and was thus amply sufficient. The upstream face was slightly stepped back whereas the width of the crest was reduced in the uppermost 4 m of the central two thirds of the dam on the downstream side. However, six buttresses assured

Large Harbaqa dam

Figure 69. Aerial view of the Harbaqa dam near Palmyra in Syria from downstream with its silted-up and partly eroded reservoir (photo De Boysson).

its stability. The dam consisted mostly of concrete poured in layers with mortar spread between them. Both faces were lined with ashlar masonry. There were three outlets at various elevations.

Huge Homs reservoir

The last dam built by the Romans in their eastern provinces was even larger than the Harbaqa dam just described above. It was located near Homs, 130 km north of Damascus in Syria (Fig. 67) (Brossé 1923). Originally attributed to the Egyptian king Sethos I (1306-1290 BC), it has meanwhile been established to date from 284 AD only (Smith 1971). It had the extraordinary length of about 2000 m and impounded approximately 90 million m³. It was thus by far the largest Roman reservoir. As usual the main body of the dam consisted of concrete lined by crudely dressed ashlar masonry on both slightly inclined faces and on the crest. Height and width changed considerably along the curved alignment. In its central part, the dam was up to 7 m high and 14 m wide at the base, i.e. grossly overdesigned. Like at Harbaqa, the crest had a reduced width still measuring 6.6 m in the uppermost 1.3 m of height. The dam was reconstructed under French occupation in 1938 and heightened by 2 m.

2.4.2 *Technology transfers*

Roman weirs in southwestern Iran

After his victory over emperor Valerian (253-260 AD) near Urfa in southeastern Turkey 260 AD, the Sassanian king Shapur I (242-272 AD) put the captured Romans to work on irrigation projects around Shush, the old Persian capital city of Susa in southwestern Iran (Fig. 67) (Graadt van Roggen 1905). Among their constructions were some superb bridges over the Karun river and two of its tributaries, which at the same time acted as diversion weirs to irrigation canals. Their principal dimensions are summarized in Table 11. As shown by the typical sections in Figure 70 the wide weirs and the bridge piers consisted of concrete and the bridge superstructures of brick work. All exterior faces were lined with excellent ashlar masonry, which also formed all the arches of the bridges. It is noteworthy that the piers were pointed in the upstream direction to break the impetus of the floods. Moreover the bridges had additional flood relief openings above the piers.

Table 11. Roman weirs in southwestern Iran.

Name of weir	Name of river	Height above river bed (m)	Length (m)	Bays no.
Ahvaz	Karun	3	900	No bridge
Dezful	Dez	3	381	22
Paipol	Karkheh	4	170	?
Shushtar	Karun	4	516	40

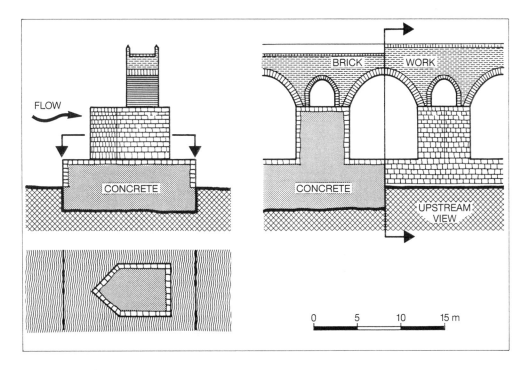

FLOW

BRICK WORK

CONCRETE

CONCRETE

UPSTREAM VIEW

0   5   10   15 m

Of all the Roman schemes in Iran the one at Shushtar was probably the finest (Fig. 71). Besides the mighty bridge-weir, the Mizan gravity dam with nine bottom outlets was built farther upstream on the inlet of the Do Dangeh irrigation canal. Later on, the latter was dammed up once more and served the operation of many water wheels. The Dariam canal started on the same river bank between the Do Dangeh canal and the bridge-weir, with a tunnel under the citadel of Shushtar. In order to execute all these construction works in the dry, the Romans first built a diversion canal of about 2 km in length on the opposite river bank. The Shushtar bridge-weir was partly destroyed by a flood as late as 1885, whereas the Dezful structure presently still serves as a bridge, despite the fact that it has lost some of its original spans (Fig. 72). Not much remains of the Ahvaz and Paipol weirs.

The Romans proceeded to a voluntary transfer of technology possibly in Armenia, which was under their domination from 64 BC to 238 (partly 384) AD. The region was endowed with a long dam building tradition, which supposedly even went back to times before the well documented oldest dams in the world at Jawa in Jordan (see above). Another predecessor was the 6.5 m high and 280 m long Dzhervezh embankment built in the second or first century BC near Abovyan, 15 km northeast of Yerevan (Fig. 67) (Agakhanian 1985). From the early fourth to the beginning of the fifth century a further three dams were built, among them the

Figure 70. Typical sections of the Roman bridge-weir combinations in southwestern Iran (after Graadt van Roggen 1905).

Shushtar weir

Dam building tradition in Armenia

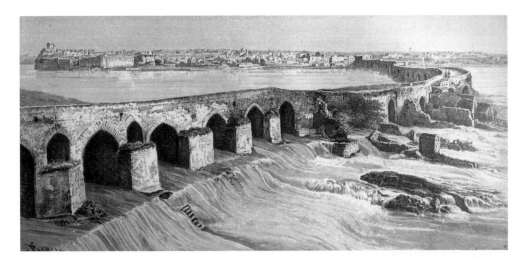

Figure 71. Painting of the Shushtar bridge-weir combination in southwesten Iran before its partial destruction by a flood in 1885.

Figure 72. The Dezful bridge-weir combination in south-western Iran; in the foreground remains of medieval mills (photo by the author).

Roman type buttress dam

Yererui buttress dam near Anipemza, 80 km northwest of Yerevan. It was 5.3 m high and 160 m long and consisted – in typical Roman fashion – of two slightly inclined masonry walls of 0.8 and 0.7 m width respectively and a 3.0 m wide earth core in-between. On the downstream face, there were 0.3 m wide and 0.7 m thick buttresses spaced 5 m apart.

2.4.3 *Byzantine epilogue*

Byzantine dams

In 395 AD the Roman empire was divided into two parts. While the western half disintegrated rather quickly under the impacts of Christianization and Germanization, the Byzantine empire in the east with Istanbul as its capital city, survived for over one millennium. The two Maagan-Michael gravity dams date back to the

beginning of this period (Peleg 1986). They impounded the Tan-
inim river sufficiently high to feed a second aqueduct to Caesarea
5 km to the south. Located in Israel 86 km north of Jerusalem, the
city was an important regional centre at the time (Fig. 67).

Whereas the northern dam was only 3.5 m high but 900 m
long, the main dam to the west was up to 7 m high and 193 m long.
Both dams had almost rectangular cross-sections with average
widths amounting to 71% of their heights. Thus, they showed
ample stability even in case of overtopping by a flood. The downstream face of the western dam was inclined 14%, whereas the
northern structure was reinforced by some downstream buttresses.
Both dams consisted of concrete lined with ashlar masonry (Fig.
73). At the left end of the western dam there was an outlet sluice
consisting of three bays. The intermediate piers pointed upstream
and, like both side walls, were provided with slots for the insertion
of planks. Later several water wheels were installed downstream of
the dam. The northern dam contained two vaulted bottom outlets.

In the sixth century AD the great Byzantine emperor Justinian
I (527-565 AD), a prolific builder, had also many hydraulic works
carried out (Procopius). Among them were the Risāfe water
supply dam in northeastern Syria, and, possibly, the Löştügün
irrigation dam near Amasya in northern Turkey, representing one
of the few pure embankments built by the Romans (Fig. 67)
(Garbrecht 1991b, Stark 1957-1958). Justinian I also ordered the
construction of several flood control dams in southeastern Turkey.
Not many remains are left of the Iron Gate dam to the east of
Antakya (Downey 1963). Likewise little is known about the rem-
nants of a dam diverting the river flowing through Urfa around the
city (Segal 1970).

Details of Maagan-Michael dams

Justinian's dam constructions

Figure 73. Downstream view of the left end of the western Maa-gan-Michael dam in northern Israel; on the right one side wall of the outlet sluice (photo by the author).

Dara curved flood
retention dam

Most interesting is, however, the dam built immediately up-stream of the fortified town of Dara, some 10 km southeast of Mardin (Fig. 67), because Procopius of Caesarea (490-562 AD) explicitly stated, that its engineer, Chryses of Alexandria, 'did not build this barrier in a straight line, but in the form of a crescent, in order that its arch, which was turned against the stream of the water, might be better able to resist its violence. The upper and lower parts of this barrier are pierced with apertures, so that, when the river suddenly rises in flood, it is forced to stop there and to flow no further with the entire weight of its stream, but passing in small quantities through these apertures, it gradually diminishes in violence and power....'. This is a late testimony, showing the Roman engineers perfectly well understood the principle of the arch dam as well as that of the flood retention basin. However, recent preliminary surveys found no trace of the arch dam, but only the remains of two gravity wing walls (Garbrecht & Vogel 1991). Nevertheless in the 70 m wide gap between them there might have been a slightly curved (central angle about 40°) polygonal structure of some 5 m of height.

CHAPTER 3

The Moslem world

3.1 ARABIA

Inspired by the new Islamic religion founded by Mohammed (570-633), the Arabs quickly developed their country and conquered, within one century, North Africa and Spain as well as southwestern Asia as far as the Indus river and Uzbekistan. Probably in continuation of the old traditions of their Yemenite predecessors, they built several irrigation dams around the new power centres of Mecca and Medina, the main data of which are summarized in Table 12 (Fahlbusch 1987).

Many dams around Mecca and Medina

All these dams were of the gravity type with two outer walls of dry masonry and an earth or rubble core in between, i.e. of a type already used one millennium earlier in Yemen and later abandoned in favour of massive gravity dams (see above). About half of the structures showed vertical walls whereas one, or, more often both walls of the others were inclined as much as 60% (Fig. 74). The strong inclinations were applied on dams with relatively wide bases, like Arda and Darwaish. About one fourth of the dams were built in two or more stages.

Structural details

Except for the 30 m high Qusaybah dam near Medina, which was slightly curved in plan, the alignments of all the other structures were straight. About one half of them were provided with a flood overflow at one end, sometimes with a downstream training wall to guide the spilled water to a safe distance from the dam's toe. Outlet works could be found in only two dams, at Agrab with four inlets to a shaft at various levels, as already applied at the Asid dam in Yemen (see above). On the other dams such installations might have been washed away when they broke. Although often silted up completely (Agrab, Sadd, Saisid and Zaydia), one third of the dams listed in Table 12 are still more or less intact. The Samallagi dam is still operative, while Khasid was reconstructed 1974-1975.

Spillways and outlets

Table 12. 7th and 8th century dams in central Arabia.

Nearest city	Name of dam	Height (m)	Length (m)	Base width (m)	Ratio width/ height
Mecca	Agrab	4.0	113	?	?
	Arda (Aradah)	5.5	315	8.0	1.45
	Dama	9.0	170	7.5	0.73
	Darwaish	10.0	150	11.5	1.10
	Luṣb	5.5	85	?	?
	Qusaybah	11.5	85	7.0	0.61
	Sa'b	2.0	150	4.0	2.00
	Sadd	6.5	50	?	?
	Saisid	8.5	58	?	?
	Salamah	5.2	52	3.2	0.62
	Samallagi (Thamalagi)	11.0	225	10.0	0.91
	Sidad (Ghumar)	3.8	110	3.5	0.92
	Thal'ba (Tha'laba)	9.0	80	10.0	1.11
	Umm al-Baqarah	5.2	63	4.2	0.81
Medina	Hashquq	2.0	130	?	?
	Khasid	6.0	40	?	?
	Qusaybah	30.0	205	?	?
	Zaydia	4.0	25	?	?

3.2 SOUTHEASTERN SPAIN

The Moslems crossed the straits of Gibraltar in 711 and stormed north through the Visigothic kingdom of Spain into France, where they were finally stopped in 732 in the battle between Tours and Poitiers by Charles Martel (715-741), grandfather of Charlemagne (768-814). Nonetheless they kept most of Spain under their domination for many centuries. In the region of Valencia, 300 km southeast of Madrid, the new rulers came from Syria, where they had taken over and developed the elaborate irrigation systems left by the Romans (Glick 1970). In the plain northwest of Valencia they could again start from the remains of Roman irrigation installations, which possibly already included diversion weirs across the Turia river. Within some 12 km of its course upstream of the city the Moslems built nine diversion weirs in the tenth century. Their dimensions are listed in Table 13 (Fernández et al. 1984).

With the exception of the oblique Moncada dam, all these weirs crossed the river perpendicularly, in a straight line. They consisted of rubble masonry laid in lime mortar and lined with ashlar blocks (Fig. 75). The widths were relatively ample, to allow for a good energy dissipation in the water flowing over the weirs at times of flood. To this end their downstream faces were either stepped

Moslems reactivate Roman irrigation

Diversion weirs

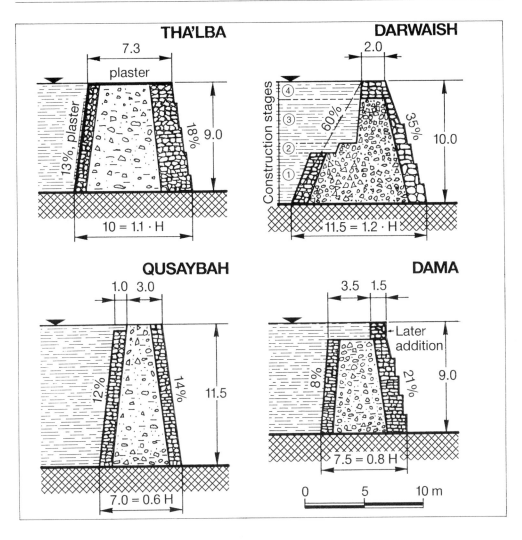

Table 13. 10th century weirs near Valencia in Spain (from up- to downstream).

Name of weir	Height (m)	Length (m)	Base width (m)	Ratio width/ height	Type of section (Fig. 75)
Depuradora	1.5	30	5.0	3.3	C
Moncada	2.0	61	9.0	4.5	A
Cuart	7.0	61	11.0	1.6	B
Tormos	2.0	56	7.0	3.5	B/C
Mislata	1.5	44	5.0	3.3	C
Mestalla	2.5	90	8.5	3.4	A
Favara	Destroyed				
Rascaña	4.0	72	10.0	2.5	B
Robella (Ruzafa)	3.5	116	8.5	2.4	C

Figure 74. Cross sections of four early Moslem dams near Mecca in Saudi Arabia (after Fahlbusch 1987).

Figure 75. Typical sections of 10th century diversion weirs near Valencia in Spain (after Fernández et al. 1984).

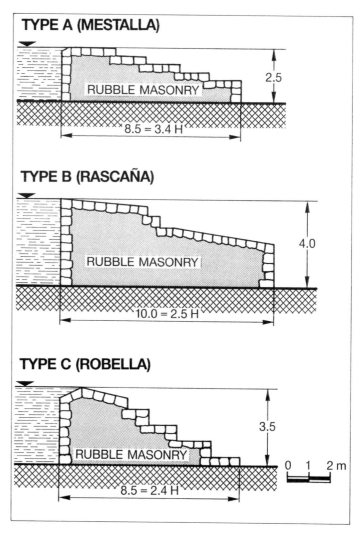

TYPE A (MESTALLA)

RUBBLE MASONRY

2.5

8.5 = 3.4 H

TYPE B (RASCAÑA)

RUBBLE MASONRY

4.0

10.0 = 2.5 H

TYPE C (ROBELLA)

RUBBLE MASONRY

3.5

8.5 = 2.4 H

0 1 2 m

(Type A in Fig. 75) or gently sloping (Type B). The overflow crest was identical with the upstream face or displaced downstream. This resulted in a roof-like section (Type C), akin to the modern parabolic cross section (see below).

Around the year 970, the large Parada weir was built on the Segura river upstream of Murcia, 350 km southeast of Madrid. The area was settled by Yemenites, who had brought along with them a millennium-old tradition in irrigation and dam engineering (see above). The up to 8 m high weir consisted of three distinct parts (Fig. 76) (Fernández et al.1984). From the left to the right river bank these were:

– a 200 m long sturdy wall ('murallón') forming the right bank of the first section of the irrigation canal;

Large Parada weir

– the 31 m long, so called 'old' weir, with a rather steep down-stream slope of 1:1.4 and, including the horizontal apron, a base width of only 15 m or 190% of the height; and

– the 74 m long, gently sloping, 'new' weir with a base width of 50 m or over six times the height.

Like the weirs near Valencia, the Parada dam was built of rubble masonry laid in lime mortar and lined with ashlar blocks. In modern times the latter were covered with concrete, which is already disintegrating (Fig. 76).

Figure 76. Downstream face of the Parada weir near Murcia in Spain; in the foreground the 'new' part; in the background the 'old' part (photo by the author).

3.3 SOUTHWESTERN ASIA

3.3.1 *Projects in various regions*

In parallel to their conquest of North Africa and Spain, the Moslems subdued southwestern Asia in the seventh and eighth centuries as far as the Indus river and Uzbekistan. Although these regions were united by a common religion and writing, they soon fell apart politically. Thus, in the ninth century the Abbasid kings in Baghdad controlled hardly more than the surrounding country-side. Therein they built a 130 m long dam on the Adhaim river, 150 km north of Baghdad (Fig. 77) (Herzfeld 1948). The 12 m high masonry structure was equally wide at its base. While the upstream face was vertical, the downstream one was stepped to 43% average inclination resulting in a 7.5 m wide crest. Along its upstream edge it had a sturdy parapet of 1.6 m height and near its

Abbasid dam on the Adhaim

Figure 77. View along the ruins of the Adhaim dam in central Iraq (photo H. Kreuzer, Dottikon/AG).

left end, a spillway with three bays of 2 m opening. The upstream heads of the piers were rounded, so as to offer a minimal resistance to the flood waters passing through the spillway.

About one century later, the local Bujid rulers had a whole series of irrigation and power dams built on the Kur river east of Shiraz in southern Iran, the dimensions of which are listed in Table 14 (Hartung & Kuros 1987). Particularly the first two of the structures were actually 'power dams' with as many as 30 water wheels being driven by the impounded water (Fig. 78). The arrangement of the wheels was that of the 'Arubah-mill', developed in the second century AD in Israel (Avistur 1960) and in use also in Iran as early as in Sassanian times near Deh Luran, 110 km northwest of Shush

Bujid power dams on the Kur

Table 14. 10th century dams on the Kur river in southern Iran (from up- to downstream).

Name of dam	Height (m)	Length (m)	No. of water wheels
Amir	9	103	30
Feizabad	7	222	22
Tilkan	6	162	8
Mawan	6	66	?
Hassanabad	Replaced by modern dam		
Djahanabad	5	50	?

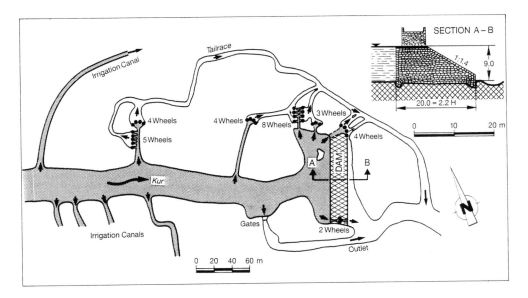

(Fig. 67) (Neely 1974). The water was led into a vertical shaft with a nozzle at its bottom, which was directed against the paddles of a horizontal wheel, much like a modern Pelton turbine. The vertical axle-tree of the wheel directly drove, one floor above the wheel chamber, the upper stone of a mill or the horizontal axis of a vertical edge-runner for crushing olives, sugar cane etc. The Amir dam itself was a simple weir of trapezoidal cross-section with a relatively steep downstream face and a base width of more than twice its height. It was built of rubble masonry laid in lime mortar. With it combined was a bridge, as encountered at the Roman bridge-weir combinations in southwestern Iran (see above) (Fig. 79).

Towards the end of the tenth century a first Turkish dynasty installed itself in Ghazni, 125 km southwest of Kabul in Afghanistan. It dominated this country for about two centuries as well as a large part of Pakistan, the east of Iran, Tajikistan and parts of Uzbekistan and Turkmenistan (Fig. 80). Under Ghaznavid rule, several large gravity dams were built in some of these countries, the

Figure 78. Plan and section of the Amir dam on the Kur river in southern Iran indicating location of the water wheels (after Reza et al. 1971).

Large Amir 'power plant'

Ghaznavid dams

Figure 79. Downstream view of Amir dam on the Kur river in southern Iran (photo by the author).

Figure 80. Map of Iran and neighbouring countries show- ing the locations of the groups of ancient Moselm dams (circles with number of dams in each group).

Dams near Samarkand

data of which are listed in Table 15 (Balland 1976, Hartung & Kuros 1987, Mukhamedjanov 1985). King Mahmud (998–1030) had three irrigation dams built near his capital city, but only of the one named after him some ruins remain in the modern Saraj reservoir, 23 km north of Ghazni (Balland 1976) (Fig. 81).

The two straight gravity dams near the ancient trading centre of Samarkand in Uzbekistan are better preserved (Mukhamedjanov 1985). While the younger Gishtbank dam was a sturdy structure, the older Khan dam, built of granite ashlar masonry, was barely stable (Fig. 82). The same was true for the Sheshteraz dam near Mashhad in northeastern Iran with an elaborate system for the

Table 15. Ghaznavid gravity dams.

Name of dam	Distance from nearest city	Height (m)	Length (m)	Base width (m)	Ratio width/height
Gishtbank	30 km SW Samarkand	~8	25	9.1	1.14
Khan	100 km N Samarkand	15.2	52	8.3	0.55
Sheshtaraz	160 km SW Mashhad	25.0	35	12.0	0.48
Soltan Mahmud	100 km SW Kabul	32.0	220	?	?

Figure 81. Downstream view of the remains of the Ghaznavid dam in the modern Saraj reservoir near Ghazni in Afghanistan (photo O. Wehle).

Figure 82. Cross sections of two Ghaznavid gravity dams (after Mukhamedjanov 1985 and Hartung & Kuros 1987).

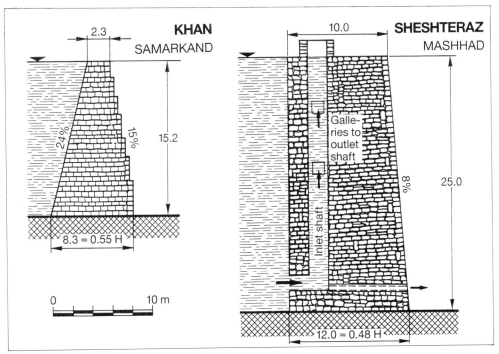

withdrawal of the water from the reservoir (Hartung & Kuros 1987). A bottom outlet led to an inlet shaft, which was connected at various elevations by short, gated galleries to a parallel outlet shaft. Floods were discharged over the dam's crest.

3.3.2 *Mongolian dams in Iran*

Mongolian conquest of Iran

In 1205 Genghis Khan (1196-1227) set out from Mongolia to conquer the world. His grandson Hulagu (1251-1265) subdued most of southwestern Asia and installed the Ilkhanid dynasty in Iran, which was islamized around 1300. Eighty years later, Tamerlane (1360-1405) reconfirmed the Mongolian conquest, but the indigenous Safavids were able to free the country of the foreign yoke in 1502. Although the Mongolians destroyed much in their first onslaught, including hydraulic installations, they soon realized their mistake and initiated the construction of several dams in Iran, on which the data are summarized in Table 16 (Hartung & Kuros 1987).

Saveh gravity dam

The first structure built at Saveh by order of Shams-ed-Din, primeminister of the Ilkhan Takudar (1281-1284), was an unfortunate one (Fig. 80). Even though its base width of 144% of the height was grossly overdesigned, the river deposits on which it was founded washed out probably upon the first impoundment and rendered the dam useless (Hartung & Kuros 1987). After having

Table 16. Mongolian dams in Iran (in chronological order).

Year of completion	Name of dam	Distance from nearest city	Type	Height (m)	Length (m)
1285	Saveh	120 km SW Tehran	Gravity	25	65
~1300	Kebar	64 km NW Kashan	Arch	26	55
~1350	Kalat	90 km N Mashhad	Gravity	26	74
	Kurit	50 km SE Tabas	Arch	60[1]	28
	Salami	200 km S Mashhad	Gravity	24	241
1400	Abbas	25 km NE Tabas	Arch	20	?
~1450	Golestan	30 km N Mashhad	Gravity	16	100
	Torogh	25 km SE Mashhad	Gravity	20	91

[1]Heightened by 4 m around 1850.

bridged the river flowing beneath it for more than 700 years, it was demolished recently to make room for a modern dam (Fig. 83).

A similar wash-out occurred at a dam 100 years younger near Tabas in eastern Iran. It was underpinned with a brick arch in the 17th century under one of the Safavid kings called Abbas, whence the dam got its misleading name (Goblot 1973). The most remarkable fact however, was that it represented the third in a series of arch dams, the first application of this type since the Romans (see above). The earliest one was built about 1300 at Kebar, 64 km northwest of Kashan in central Iran (Fig. 80) (Goblot 1965). The middle third of the 26 m high and 55 m long structure was curved with a radius of 35 m and a central angle of only 40° (Fig. 84).

In the upper half of the dam the arch did not abut against the canyon walls, but as in the dam of Monte Novo in southeastern Portugal, probably of Roman origin (see above), it passed on both sides into straight wing walls. With a base width of 58% of their height these walls were stable by themselves. At midheight the arch was stressed to some 0.6 MPa in compression and 0.3 MPa in tension. This was acceptable in view of the good quality of the limestone masonry laid in lime mortar containing ash from thorn bushes. These gave the lime hydraulic properties comparable to volcanic ash. Like some ancient dams and all the southwestasian ones since the 11th century, Kebar was provided with a shaft near the upstream face, which contained a spiral staircase and had several inlets at various elevations as well as an outlet to the irrigation canal at the bottom.

The third Mongolian arch dam is especially remarkable for its extraordinary height of 60 m (Goblot 1973). After a heightening

Abbas arch dam

Kebar arch dam

Structural details

Outlet works

Figure 83. Upstream view of the impressive Saveh gravity dam in Iran; its foundation was probably washed out upon its first impoundment (photo by the author).

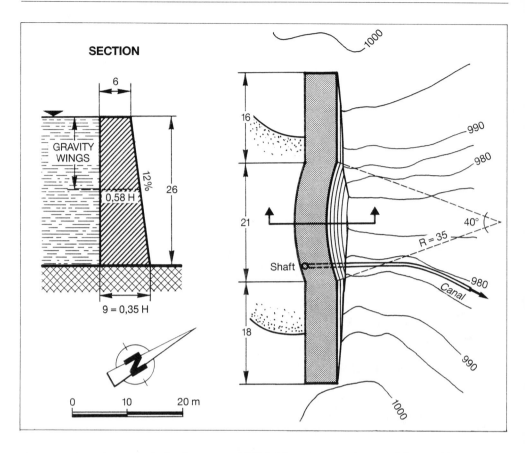

Figure 84. Plan and section of the Kebar arch dam in central Iran (after Hartung & Kuros 1987).

Record high Kurit dam

Mongolian gravity dams

by 4 m around 1850 this was to remain a world record for any type of dam, until early in the 20th century. Similarly to the aforementioned Abbas dam, the Kurit dam was built in a very narrow canyon and its crest length reached only 44% of the height (Fig. 85). It therefore remained erect, despite the fact that a large segment of the vertical downstream face broke away in its lower half. In the process, a shaft and other openings were exposed as well as the same type of masonry found at the Kebar dam.

Not much is known about the Mongolian gravity structures built after the Saveh dam (Table 16) (Hartung & Kuros 1987). Like this one they were extremely sturdy structures with large crest and base widths, the latter usually exceeding the height. Since floods were always spilled over the dam crest, the downstream face of most gravity structures was inclined or stepped as in weirs. Nevertheless, many dams were damaged and, over the centuries, some even breached due to either erosion of the downstream face or scouring in the river bed at the toe. As for the dams which survived, the reservoirs gradually silted up because the small inlets to the outlet shafts had practically no purging effect.

Figure 85. Crest of the Kurit arch dam in eastern Iran ultimately 64 m high (photo R. Carlin).

3.3.3 *Post-Mongolian epilogue*

After the indigenous Safavid dynasty took over in Iran, dam building went on, especially under king Abbas II (1642-1667). However, the construction of arch dams was discontinued and only gravity dams were erected, mostly of the aforementioned sturdy type. Their data are listed in Table 17 (Hartung & Kuros 1987, Mukhamedjanov 1985).

Most of the dams for which adequate information is available are situated in northeastern Iran and Uzbekistan. Here, the Abdullakhan dam was built in as early as the 16th century (Mukhamedjanov 1985). Its width slightly exceeded the height and its downstream face was stepped to an average inclination of 69%. Its outlet structure showed two novel features. Firstly, the inlets to the outlet shaft were replaced by a continuous slot into which narrow gates could be inserted, one on top of the other. Secondly, the gallery leading from the bottom of the shaft to the downstream toe was

Safavid dams

Table 17. 16th-17th century dams in southwestern Asia.

Name of dam	Distance from nearest city	Height (m)	Length (m)
Abdullakhan	150 km NW Samarkand	15.0	85
Akhlamad	82 km SE Mashhad	14.3	220
Fariman	90 km SE Mashhad	21.0[1]	90
Ghorud	38 km SW Kashan	23.0	100
Kerat	200 km S Mashhad	6.0	150
Khadjoo	335 km S Tehran	3.0	180

[1]Heightened by 1.5 m in the 1930's.

Figure 86. Downstream view of the Khadjoo bridge-weir combination in Isfahan in Iran (photo by the author).

Beautiful Khadjoo weir

prolonged to the upstream face and closed with a gate. For the purging of sediments this gate could be opened when the reservoir was almost empty. Moreover, the Abdullakhan dam had a small spillway chute near its right abutment.

Whereas the Akhlamad dam was a very sturdy structure, the nearby Fariman dam, built under king Abbas II, had a base width of only 69% of its height (Hartung & Kuros 1987). It was strengthened in the middle of the downstream face some 150 years later by means of the addition of a massive pyramid of 17 by 25 m base area. The last dam in Table 17 was a bridge-weir combination also built under Abbas II, akin to those erected in southwestern Iran by Roman prisoners of war some 1400 years earlier (see above). This Khadjoo weir in the city of Isfahan is certainly one of the most beautiful hydraulic works not only in the Moslem, but also in the entire world (Fig. 86)!

CHAPTER 4

Medieval eastern Asia

4.1 SRI LANKA

In the preceding chapter dealing with the construction of dams by the ancient civilizations, the description of the developments on the island of Sri Lanka (formerly Ceylon) at the southern tip of India was interrupted at about the end of the classical antiquity in the fifth century AD. Although this had no repercussion on Sri Lanka, it coincided with the completion of the first large water management system for the capital city of Anuradhapura in the northwest of the country (Fig. 31). Seventy-five km to the southeast, in the basin of Sri Lanka's largest river Mahaweli, Polonnaruwa became the new capital of the Singhalese kings in 781 AD.

Completion of Anuradhapura system

Long before this date several water management projects had already been constructed in the Mahaweli basin (Fig. 87) (Brohier 1934-1935, Fernando 1982). The first ones were isolated reservoirs like Allai and Horabora in 170 and 160 BC respectively. The earliest system in 70 AD was the 34 km long canal from the Elahera weir on the Amban river to the Minneriya reservoir. This storage basin was expanded to a capacity of 136 million m^3 in 290. Together with the Kantalai reservoir of 610 it represents one of the largest impoundments of ancient Sri Lanka. In 370 another canal was built from Angamedilla on the lower Amban to fill the Topa reservoir near Pollonaruwa. In order to supplement the flow in this river a canal of 80 km was constructed between 470 and 680 from the Minipe weir on the Mahaweli to Angamedilla. Moreover this goal was complimented by a shorter canal built in 610 from Hattote on the Kalu to Elahera. These canals also served the dual purpose of irrigating the areas they crossed and of inland navigation.

Development of Mahaweli basin

The described works set the stage for the enlargement of the Topa reservoir under king Parakrama (= the Great) Bahu (1153-1186) to his 'samudra' (= sea) at Polonnaruwa, which could store 102 million m^3 of water. Contrary to popular belief it was thus not

Parakrama's 'sea'

Figure 87. Map of the reservoirs and diversion canals (broken lines) in the Mahaweli river basin on Sri Lanka; figures in frames indicate the reservoir capacity in millions of m³ (for reconstructed reservoirs present values) (after Brohier 1934–1935).

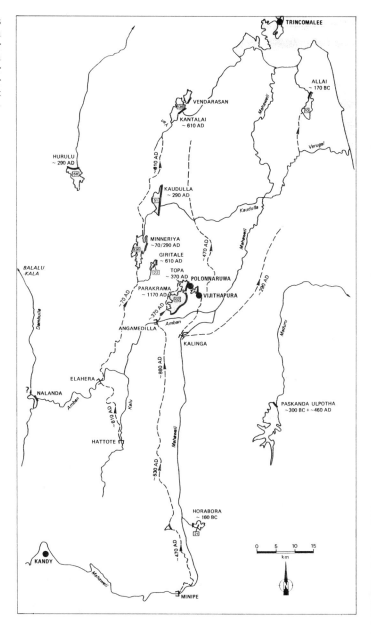

the largest reservoir in ancient Sri Lanka although it was impounded by the largest embankment built so far. With a maximum height of 15 m and a crest length of 13 600 m it required 4.6 million m³ of fill, a world record until 1912 (Gatun dam in Panama). Another big scheme, begun under Parakrama Bahu in the north-

'Giant's tank'   west of Sri Lanka, included the Sodiyon Kattu Karei, i.e. 'the giant built embankment', 52 km west of Vavuniya (Brohier 1934–

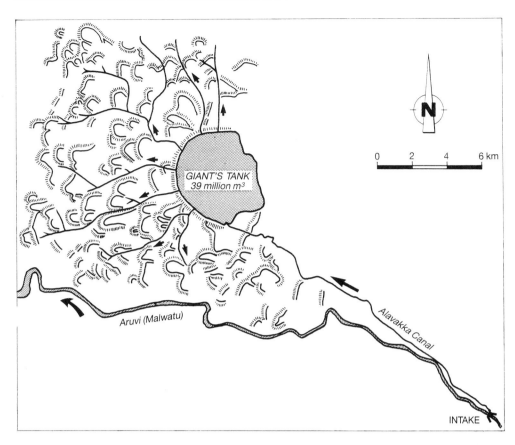

GIANT'S TANK
39 million m³

0   2   4   6 km

Aruvi (Malwatu)

Alavakka Canal

INTAKE

1935). It was some 9 km long and semicircular in plan (Fig. 88). The storage capacity amounted to 39 million m³.

Besides these two large reservoirs Parakrama Bahu boasted to have built some 1800 smaller new dams and to have restored some 2300. This is probably grossly exaggerated, but the 'tyrant megalomaniac' (Leach 1959) was also addicted to military conquest and embellished his capital with many new buildings, thereby literally bleeding his people to death. He was followed by chaos and under the increasing pressure of Tamil invaders from southern India, the Singhalese reign collapsed. Thus its magnificent hydraulic works lost their socioeconomic basis and fell into disrepair and oblivion. Only in the last quarter of the 19th century reconstruction of the relatively short breaches in the embankments began and the utilization of the ancient reservoirs was resumed under British occupation.

Figure 88. Plan of the uncompleted Giant's Tank scheme in northwestern Sri Lanka and the older farmers' ponds (after Brohier 1934-1935).

Collapse of Singhalese reign

4.2 SOUTH INDIA

At about the same time the Tamil invaders let the hydraulic works on Sri Lanka go to ruin (see above), there was a considerable

Dam building boom in
South India

upsurge in dam building in their southindian home country. This was dominated by various Hindu dynasties (Pallava, Chola, Vijayanagara etc.), with a short intermission of Moslem rule (Sultanate of Delhi) around 1300. The data of dams built before and, for the most part, after this event are summarized in Table 18 (Rao 1951 and 1961). The oldest one listed already showed characteristics similar to the ones of the ancient Singhalese dams (see above), i.e. a moderate height and a very long crest. The large Veeranam reservoir was filled not only by the runoff from the natural catchment basin of 427 km^2, but was fed also by a diversion canal from the Coleroon river to the south. This canal as well as the one emanating from the reservoir were additionally used for inland navigation like in Sri Lanka, linking the coast and the city of Tiruchirapalli, 300 km southwest of Madras.

Very long Veeranam
embankment

In contrast to the Singhalese dams the embankments in southern India had considerably steeper outer slopes because the soils used in their construction were mostly gravelly. An extreme example was the Motitalav dam north of the city of Mysore, with over 24 m the highest of them all. Its upstream face was protected by ashlar blocks and had a 1:1.5 slope, whereas its downstream face sloped 1:1 (Fig. 89). In order to obtain sufficient mass to resist the water pressure, its crest width exceeded the dam's height. Such a design was applied in the same period for two dams of 26 and 12 m

Motitalav dam

Table 18. Medieval embankments in southern India.

Year of com- pletion	Name of dam	Distance from Madras (km)	Height (m)	Length (m)	Slopes Up- stream	Down- stream	Reservoir capacity (million m^3)
1037	Veeranam	210 S	*9.1*	*16 100*	1:1.5	1:1.5	*51*
11th cent.	Malkapur	490 NW	*11.6*	3780	1:2.0	1:3.0	?
11th cent.	Motitalav	370 W	*24.4*	4023	1:1.5	1:1.0	22
1213	Lakhnawaram	560 N	15.2	610	1:2.5	1:2.5	*59*
1213	Pakhal	540 N	19.2	1220	1:2.5	1:2.0	*96*
1213	Ramappa	570 N	17.1	610	1:2.5	1:3.0	*153*
1369	Anantha	190 NW	14.3	?	1:2.0	1:2.0	21
1369	Anantharaja (Porumamilla)	230 NW	11.0	1370[1]	1:1.5	1:2.0	20
14th cent.	Chembarambakam	26 SW	9.8	8850	1:1.75	1:2.0	88
14th cent.	Cumbum	310 NW	18.3[2]	296[2]	1:1.5	1:2.0	105[2]
14th cent.	Rasool	?	12.8	1280	1:1.5	1:2.0	?
1400	Kaveripakkam	90 W	12.1	7240	1:1.5	1:2.0	40
15th cent.	Kesari	?	12.2	1520	1:2.0	1:2.0	?
1520	Peddatippa	200 W	16.0	1265	1:2.0	1:3.0	12
16th cent.	Ibrahimpatan	490 NW	15.2	2410	1:3.0	1:2.0	?

[1]Three sections between four hills. [2]Data from placard on site.
Italics: Record breaking dimensions.

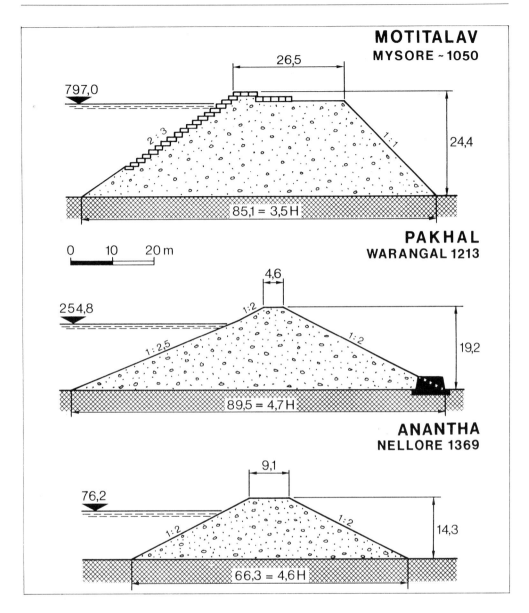

MOTITALAV
MYSORE ~ 1050

26,5

797,0

2 : 3

1 : 1

24,4

85,1 = 3,5 H

0   10   20 m

PAKHAL
WARANGAL 1213

4,6

254,8

1:2

1 : 2

19,2

1 : 2,5

89,5 = 4,7 H

ANANTHA
NELLORE 1369

9,1

76,2

1 : 2

1 : 2

14,3

66,3 = 4,6 H

height as well as 92 and 460 m length respectively. They contained the enormous Bhojpur lake near Bhopal, some 600 km south of Delhi (Kincaid 1888). However, the higher of the two dams was breached only three centuries later and the reservoir area of 650 km² reclaimed for cultivation. The last dam with an excessive crest width of 20 m was the 18 m high Cumbum embankment built in the 14th century (Fig. 90).

Quite modern looking cross sections were adopted already around 1213 for the Lakhnawaram, Pakhal and Ramappa dams

Figure 89. Cross sections of three South-Indian embankment dams (after Rao 1961).

Figure 90. Downstream view of the Cumbum dam in southern India; foreground shows the spillway chute (photo H. Kreuzer, Dottikon/AG).

Modern looking cross sections

Requisites for a good dam

Faults to be avoided

east of Warangal, some 550 km north of Madras (Fig. 89). And near the Anantharaja (or Porumamilla) dam, farther south, an inscription dated 1369 enumerates the following – still valid – twelve essential requisites for the construction of a good reservoir (Vadera 1965):

1. A king (i.e. owner or client) endowed with righteousness, rich, happy and desirous of acquiring fame;

2. A person well versed in hydrology;

3. A reservoir bed of hard soil;

4. A river conveying sweet water from a distance of about 40 km;

5. Two projecting portions of hills in contact with the river;

6. Between these projecting portions of hills a dam built of compact stone, not too long but firm;

7. The two extremities of the hills to be devoid of fruit-bearing land (i.e. humus);

8. The bed of the reservoir to be extensive and deep;

9. A quarry containing straight and long stones;

10. Fertile low and level (i.e. irrigable) area in the neighbourhood;

11. A watercourse having strong eddies in the mountain region; and

12. A group of men skilled in the art of dam construction.

The inscription also enjoined that the following six faults should be avoided:

1. Oozing of water from the dam;

2. Saline soil;

3. Site at the boundary of two kingdoms;

4. High ground in the middle of the reservoir;

5. Scanty water supply and an extensive area to be irrigated; and

6. Too little land to be irrigated and excessive supply of water.

Finally the inscription stated that 1000 people with 100 carts worked for two years on the construction of the Anantharaja dam (RAO 1951).

According to the above and/or similar rules, several 10 000 reservoirs were built in southern India in the course of the centuries. Then, the 16th century saw the end of the Hindu state of Vijayanagara and the arrival of the first European conquerors who installed themselves at various points on both coasts. As early as around 1520, one of the last Vijayanagara kings sought the advice of the Portuguese engineer Joao de la Ponta in the construction of a reservoir for his new city of Nagalpura (Shrava 1951). Thus began the transfer of European dam technology to India, which was to become especially important during the two centuries of British rule over the subcontinent (see below).

Introduction of European technology

4.3 CAMBODIA

Situated between India and China, Cambodia became involved in the increasing inter-regional trade early in the first millennium AD and the country was thereby quickly indianized (Liere 1980, Moore 1989). In this process the Cambodians undoubtedly became aware of the ancient irrigation reservoirs in Sri Lanka (see above). At that time, or, perhaps even before, they surrounded their cities in northern Cambodia and adjoining Thailand often with multiple, circular moats filled with water. These not only protected the settlements and their temples physically and spiritually but they were also used for water supply and irrigation during the dry season.

Early city moats

Under the Khmer god-kings, from the ninth century onwards, these moats were arranged rectangularly and strictly oriented in north-south as well as east-west directions, thus projecting heaven onto earth, according to the Indian cosmology. In their centre was the sacred mount as seat of the divine king surrounded by the quarters of a large priesthood and its assistants. The moats were complemented by huge rectangular basins, in analogy to those near the still famous capital city of Angkor, 210 km northwest of Phnom Penh (Fig. 91). They covered up to 17 km^2, were 3 to 5 m deep and capable of storing up to 70 million m^3 of water. The excavated material was not only used in the earth dikes around the basins, but also to build the abovementioned temple mounts. The basins neither had outlets nor did they feed any water distribution systems. Consequently they had only marginal significance for irrigation.

Rectangular Khmer moats and basins

Angkor system

For a crude regulation of the flow into the basins, some 20 flood retention dams were erected on the headwaters of the Siemrap

Figure 91. Map of the basins near Angkor in northern Cambodia (after Moore 1989).

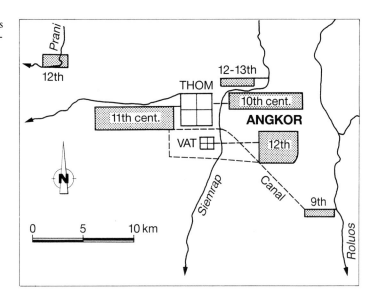

Flood control

river 30 km northeast of Angkor. They were also aligned either north-south or east-west and consisted of homogeneous embankments with 1:5 slopes on both faces. Their outlets had no gates and their reservoir volumes of approximately 4 million m³ added up to only a small fraction of the content of the basins. A true reservoir had been built for the supply of the moats and temples of the former capital city of Işanapura, 130 km southeast of Angkor. There was also a flood retention dam nearby. After the 13th century, the theocratic superstructure of the Khmer largely collapsed although the profane basic structures of the Cambodian society continued to

Collapse of Khmer reign

exist, thus perpetuating many traditional technologies to the present day. In the 1960's, when the huge 11th century basin west of Angkor was equipped with outlets and a distribution system, not a single drop of water was used to irrigate the paddy fields of the project area (Liere 1980)!

4.4 CHINA

As in ancient Greece (see above), the abuse of dams for military purposes was resumed during the division of China in the years between 317 and 589 (Zheng Liandi 1991). The tactics of

Bellicose abuse of dams

'shuigong' (= against by water), i.e. the flooding of enemy positions by building or breaching dams and dikes, was used in some twenty instances, the most spectacular of which occurred in 514/15 on the Huai river near Fushan in central China (Fig. 40). There, a 32 m high and around 4000 m long embankment of over 2 million m³ of volume was built in order to flood an enemy garrison farther upstream. Closure of the dam proved to be a difficult enterprise and succeeded only in a second attempt when stone filled timber cribs and tens of thousands of tons of iron pieces were thrown into the river. But a mere four months later the mighty Huai river overtopped the dam despite two spillways. The reservoir content of 10 000 million m³ of water released by the breached dam caused havoc in the underlying area and killed over 10 000 of its inhabitants.

Fushan embankment

After its reunion under emperor Weng (580-604), China quickly reached a new peak in its long history. During this period and the following centuries two weirs were reportedly constructed at Tongji and Tashan (in 833) south of Shanghai (Fig. 40) (Zheng Liandi 1991). Whereas the second, a 9 m high and 134 m long structure, is still in operation, the one century younger Zhengpiqu weir of 150 m length in the northwest has been abandoned (CHINCOLD 1987). The 7 m high and 232 m long Mulan weir, built in the 11th century near the mouth of the Mulanxi river on the southeastern coast is still in working order. It was to prevent the backflow of the tide and to deviate part of the river discharge for irrigation (Fig. 92) (Zheng Liandi 1991).

Several weirs

For the foundation of the masonry structure, a 12 m deep pit of equal width and length had to be excavated in one place. It was filled with stones linked by crotchets. The dam consisted of a massive northern half and a spillway to the south (Fig. 93). The latter had originally 32 and later 28 bays of 1.9 to 3.1 m width, which could be closed by means of boards placed in appropriate slots in the piers. For these, 1.5 m thick and 4.0 m wide blocks were piled up to above the water level. There was a desilting sluice at the weir's southern end near the inlet to one of the irrigation canals, whereas the intake of the northern canal was arranged well up-

Details of Mulan weir

Figure 92. Downstream view of the Mulan weir on the south-eastern coast of China (photo Zheng Liandi, Beijing).

Figure 93. Plan of the Mulan
weir (after Zheng Liandi 1991).

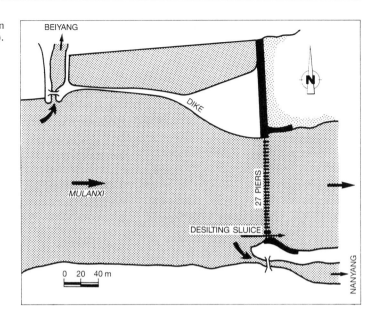

stream of the weir. It was kept free of silt deposits by a dike in front
of the massive section directing the river flow towards the spillway.
Both downstream banks were protected by stone work.

More weirs were built in the following centuries in the north at
Yinqui (1261), Gangchen and Daicun (1411), the latter being
1600 m long and still in operation today (CHINCOLD 1987,

Huge Hongze reservoir   Zheng Liandi 1991). Some 100 years before China fell again under
Mongolian domination to stay for a long time (Manchus 1644–
1912), it began the construction of its largest reservoir (Zheng
Liandi 1991). Actually, the capacity of 13 000 million m³ of the
Hongze (formerly Gaojiayan) lake in central China constituted a
world record up to the impoundment of the three times larger
reservoir behind Hoover (formerly Boulder) dam in the USA in
1935.

There were flood protection dikes at the origin of the Hongze
dam. These had to be built after the Huang (Yellow) river altered
its final course to the south and into the mouth of the Huai river in
1324, thus hindering the latter's discharge (Fig. 40). The Hongze
lake area on the Huai therefore became a flood retention zone, so
Development of   that it was decided to transform it into a permanent water body
Hongze dam   sometime before 1553, the year in which the Hongze dam was
recorded for the first time. A systematic, over 30 km long embankment was built in 1578 under the direction of Pan Jixun, a famous
expert in river training in his time. However it only dammed the
northern half of the eastern lake shore, while the southern half
acted as a spillway into Lake Baima (top of Fig. 94).

This section was closed later on, thus increasing the length of the

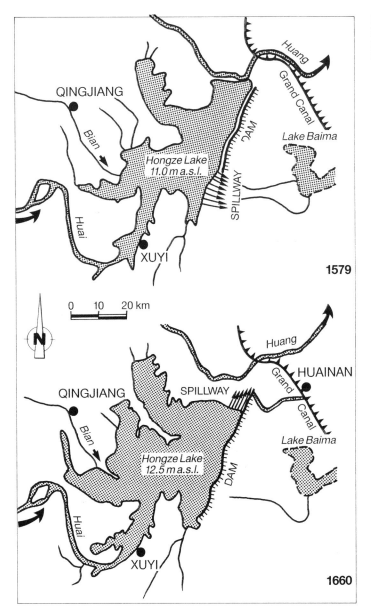

QINGJIANG

Bian

Hongze Lake
11.0 m a.s.l.

Huai

XUYI

Huang

Grand Canal

Lake Baima

DAM

SPILLWAY

1579

0 10 20 km

N

QINGJIANG

Bian

SPILLWAY

Hongze Lake
12.5 m a.s.l.

DAM

Huai

XUYI

Huang

HUAINAN

Grand Canal

Lake Baima

1660

Figure 94. Maps of the Hongze lake in central China in 1579 (top) and 1660 (bottom) (after Zheng Liandi 1991).

Hongze dam to over 67 km and raising its storage water level by 1.5 m. The Huai floods were now spilled to the north into the Huang and helped to scour away part of its notorious sediment load (bottom of Fig. 94). The dam, up to 7 m high, consisted of an 0.8 to 1.2 m wide ashlar masonry wall linked by iron cramps and founded on piles. It was backed-up by an earthfill. First there was a wall of bricks between the masonry wall and the fill and, down-stream, a mixture of clay, sand and lime. The latter as well as the

Details of Hongze embankment

mortar of the masonry and the brickwork contained glutinous rice soup as a binding agent.

4.5 JAPAN

Origins of dam building

Unfortunately little has been published in a western language concerning the development of dam building in Japan, although some 80 such structures of over 8 m in height were built between the fourth and eighth centuries AD (Fukuda 1991, JAPCOLD 1984). Their construction became possible after the unification of the country around 300 AD by the rulers of Yamato near. Osaka, 400 km southwest of Tokyo. With the help of Korean scientists and engineers many irrigation dams were constructed for rice cultivation around the capital city. The Sayama embankment, 20 km south of Osaka, completed in 380 to a height of 8 m and a crest length of some 900 m still exists. It was heightened to 10 m in the eighth century, in parallel with the construction in 703 of the 10 m and some 80 m long Mannō dam on the northeastern coast of Shikoku island, 120 km southwest of Osaka. Both dams were again heightened later on, so that little is left of the original structures.

Sayama and Mannō embankments

Details of dams

Aside from the 80 structures of over 8 m of height, many smaller ones were built. All were homogeneous embankments with rather steep slopes of 1:1.0 to 1.5. They also had wide crests, the width of which equalled or exceeded the dam height. There were overflows and canals for the release of excess water at one or both ends of the crest. All dams originally had wooden and, later on, stone outlets through their base, in which the flow could be controlled. The embankments were usually built during the short intermission from September to November between two rice growing seasons by the local peasantry, very often under the direction of Buddhist priests. Thus, in the eighth century the priest Gyoki became famous for his teaching of dam construction techniques whereas a monk named Kōbōdaishi instructed the farmers in the repair and heightening of the abovementioned Mannō dam, after it had been breached by a flood in 814. In the following centuries these procedures were maintained, although the largest embankment height achieved increased rather slowly to 15 m in the 12th century and to above 20 m in the 17th century. After the forcible opening of Japan by commodore Matthew C. Perry (1794-1858) in 1853, it adopted western technology with awesome success.

Direction by priests

Adoption of western technology

CHAPTER 5

Medieval and postmedieval Europe

It took Europe several centuries to recover from the impacts of Christianization and Germanization. In the eighth century, i.e. approximately 300 years after the disintegration of the western half of the Roman empire, the Franks tried to revive it, followed by the German emperors in the high Middle Ages. At the same time a novel, almost entirely affirmative attitude towards technology developed, which led to the improvement of ancient devices, many new inventions as well as their rapid diffusion throughout western and central Europe (White 1978).

European recovery

5.1 POWER DAMS

5.1.1 *Medieval power dams*

One technology, which spread almost explosively in the 11th and 12th centuries, was the use of water power. The water wheel had already been invented in the first century BC – interestingly enough in China at about the same time as in the West – but it encountered only a few applications, mainly to grind cereals. The invention of the cam in the ninth or tenth century triggered off a wide diversification of applications of vertical as well as horizontal water wheels: for stampers, graters, hammers, presses, saws, bellows, pumps etc. (Reynolds 1983). Concurrently the water wheels were moved away from the rivers and their dangerous floods, ice-jams and floating debris. Preference was given to the diversion of the required discharge by means of a canal or channel to the water wheel. This necessitated a diversion weir in the river, which often was amplified into a storage dam. Reservoirs ensuring a steady supply of water were also impounded in lateral valleys or excavated along the diversion canal.

Spread of water power

Diversion and storage dams

The earliest medieval diversion dams in Europe consisted of two rows of posts interconnected by lattice work and interfilled

Temporary weirs

Mill ponds

First power dams in Ore Mountains

Cento dam

with earth and stones, not unlike the Maravilla and Nezahualcoy-otl dams in Precolumbian Mexico, described earlier. These structures were of course of temporary nature and had to be repaired or even rebuilt after every major flood. One of the largest dams of this type was the 400 m long Bazacle weir built before 1170 obliquely across the Garonne river in Toulouse in southwestern France (Reynolds 1983). But at the same time the impoundment of mill ponds by embankments began and, rather rarely, by masonry dams. The few of which sufficient data are available, are listed in Table 19 (Beier 1983, García-Diego 1977, Reynolds 1983, Schnitter 1992, Smith 1971, Wagenbreth 1991). Actually, hundreds of such dams were built up to the end of the 15th century.

Among the embankments listed in Table 19 those of Greifen (formely Geyer) and Filz in the Ore Mountains south of Leipzig in Germany are remarkable, because they were the first ones built in connection with the medieval use of water wheels in mines (Wagenbreth 1991). These drove either bucket-chains and, since about 1530, series of suction pumps for dewatering or hoists for the removal of the ore. For the latter application the wheels were partitioned into two parts with the buckets inclined in opposite directions. By changing the water flow from one part to the other, the direction of rotation of the wheel and the attached windlass could be reversed. The storage reservoirs were sometimes fed by intricate systems of water collecting canals of several km of length.

Of the two masonry dams shown in Table 19 the one under the name of Cento in northern Italy was an overflown weir with a vertical upstream and a gently sloping downstream face (Smith 1971). Its particularity consisted of the fact that it was made of bricks and wood. The mill house was situated at one end of the weir and supplied directly with water from the impoundment, an arrangement very similar to modern low-head power plants. Equally modern was the location of the mill house near the toe of the 19 m

Table 19. Medieval power dams in Europe (of which dimensions are known).

Year of completion	Name of dam	Distance in km from city/country	Type	Height (m)	Length (m)	Reservoir capacity (million m³)
Ca. 1150	Lucelle	30 SW Basle/Switz.	Embankment	4	70	?
Before 1170	Bazacle	Toulouse/France	Crib	?	400	?
1404	Greifen (Geyer)	90 SE Leipzig/Germ.	Embankment	4	?	0.06
1424	Biessenhofen	16 NW St. Gall/Switz.	Embankment	4	150	0.14
Ca. 1450	Cento	80 SE Bologna/Italy	Gravity	6	71	?
1460	Bommer	28 NW St. Gall/Switz.	Embankment	5	100	?
1485	Filz	85 S Leipzig/Germ.	Embankment	6	240	0.10
1500	Castellar	120 NW Sevilla/Spain	Buttress	19	100	0.30
1500	Kornthan	100 NE Nürnberg/Germ.	Embankment	5	?	0.25

high Castellar storage dam in southwestern Spain (Fig. 95) (García-Diego 1977). It contained three horizontal water wheels to which the water was led by an equal number of vertical shafts within the dam. These shafts acted also as water intakes, like near the tenth century dams on the Kur river in southern Iran (see above). The cross section of the dam was almost rectangular with a width of only 36% of the height. Thus the sturdy end and intermediate walls of the mill house were actually buttresses ensuring the dam's stability. It certainly is no coincidence that this structural concept re-emerged in a region in which the Romans had built many buttress dams (see above).

Figure 95. Downstream view of the Castellar dam and millhouse in southwestern Spain, now submerged by a modern reservoir (photo by the author).

Castellar dam

5.1.2 *Postmedieval power dams for mining*

In the abovementioned mining area in the German Ore Mountains many more storage embankments were built in postmedieval times, whereby impermeabilization by a central clay core was used for the first time for a modern embankment in 1558 at the 9 m high Berthelsdorfer dam, 85 km southeast of Leipzig (Sieber 1992). For the mines in the German Harz Mountains some 70 km southeast of Hannover the use of water power was introduced in 1535 by immigrants from the Ore Mountains (Fig. 96, Table 20) (Schmidt 1989). Within the following three centuries around 450 to 500 km of canals were built as well as over 110 storage reservoirs. However, they were not all in operation simultaneously.

First embankments with clay core

Harz dams

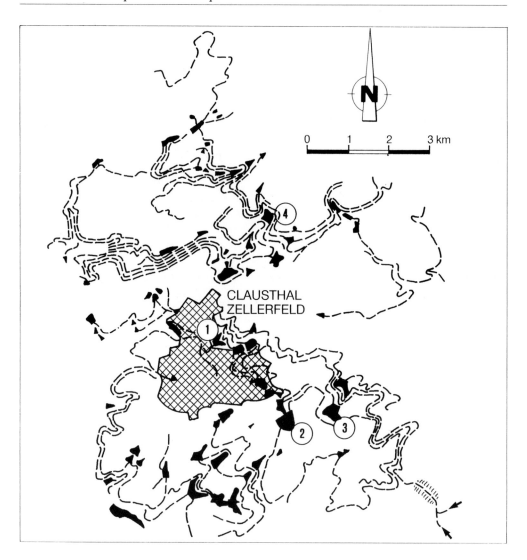

Table 20. Important power embankments in the Harz mountains in Germany.

Year of completion	Name of dam	Location of dam (No. in Fig. 96 or distance in km)	Height (m)	Length (m)	Reservoir capacity (million m^3)
Before 1551	Mid. Eschenbach	No. 1	11	119	0.06
Ca. 1560	Herzberg	3 S Goslar	12	140	?
1660	Hirschler	No. 2	11	414[1]	0.70
Ca. 1670	Jägersbleeke	No. 3	14[2]	222	0.41
1715	Wiesenbeeke	33 S Goslar	15	149	?
1718	Kupferrose	28 S Goslar	15	115	?
1722	Oder	18 SE Goslaar	22[1]	166	1.67[1]
1724	Mid. Kellerhals	No. 4	15	279	0.47

[1]Record for this group of dams. [2]After two heightenings by 1.5 m in 1697 and 3.0 m in 1719.

Some 90% of the storage dams were homogeneous embankments with similar and steep slopes on both sides (top of Fig. 97). For impermeabilization the upstream faces were lined with turfs arranged in horizontal layers 0.9 to 1.2 m wide at the dam's crest and 1.9 to 2.4 m wide at its heel. Here the turf lining was prolongated through the overburden down to sound rock. Construction

See Page 110

Figure 96. Map of the region of Clausthal-Zellerfeld in the German Harz Mountains showing the mining reservoirs and part of the canals (after Schmidt 1989).

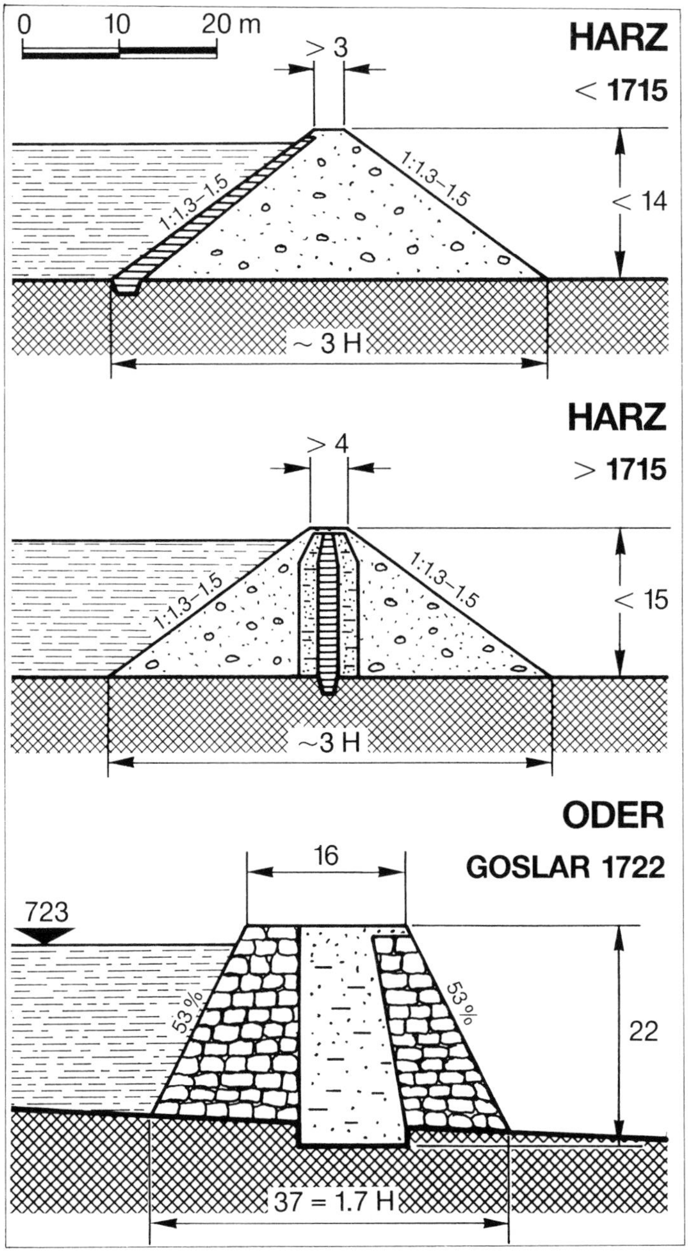

Figure 97. Standard profiles of the old (top) and new (middle) embankment types in the German Harz Mountains; bottom: cross section of the Oder dam (after Schmidt 1989).

**Embankment
construction**

was, like in Roman times (see above), by pick and shovel, with the wheelbarrow now added for transportation. The construction materials were obtained in the vicinity of the dam and were not compacted after placement. Therefore, the embankments settled considerably after their completion and had to be brought back to their original height by additional fill. In one third of the cases this operation was combined with a heightening of the embankment to cope with the ever increasing need for more storage capacity.

All dams were provided with overflows at one end of their crest. These were from 2 to 10 m long and 1 to 1.5 m deep. They were originally built of wood and later replaced by masonry structures. In view of the rudimentary hydrological knowledge of the time, the selection of the proper spillway discharge capacity was problematic. About 10% of all dams failed at one time or another, mostly due to overtopping by floods. Naturally, all dams had an outlet through their base consisting of one or more trunks into which a channel of 0.24 by 0.24 m section had been cut and covered with a plank. At their upstream ends these outlet 'pipes' had an opening which could be closed with a plug operated by means of a pole from a small bridge projecting into the reservoir. Sometimes the pole and thus the plug could be operated by means of a lever directly from the crest of the dam.

Spillways and outlets

**Impervious cores in
Harz embankments**

The upstream turf lining of the embankments was highly susceptible to damages by ice or waves and needed considerable maintenance and repair. Therefore, starting with the Wiesenbeeke dam completed in 1715, the turf was moved to the centre of the embankment and supplemented on both sides with transition zones of the sandy clay resulting from rock weathering, thus adopting an impervious core as used in the Ore Mountains since 1558 (middle of Fig. 97). Along with the impermeabilization, the plug on the outlet and its operating pole were moved to a shaft in the dam's centre which was connected through a similar pipe to the reservoir. It therefore had to be sealed with turf as well.

Oder dam

The central core concept was also adopted for the Oder dam, built from 1715 to 1722 (bottom of Fig. 97, Table 20). Since neither earth nor turf were available nearby, boulders and sandy clay resulting from rock weathering were used at the suggestion of the mining official Caspar Dannenberger (before 1650-1713) (Fig. 98). The detailed design was prepared by Bernhard Ripking (1682-1719), a young mining engineer who had traveled widely and seen some rockfill dams in Sweden. He provided two steeply inclined outer shells of dry boulder masonry resulting in a gravity dam with an earth core akin to the ones used in ancient times among the Mycenaeans, Urartians, Sabeans, Nabateans and Baluchis as well as occasionally adopted by the Romans and Moslems (see above). Although the water pressure acted directly against the

Figure 98. Upstream view of the Oder dam in the German Harz Mountains; in the foreground its construction materials: boulders and sandy clay resulting from rock weathering (photo by the author).

core, the remaining downstream part with a base width of 22 m (or 100% of the height) was stable. The upstream shell was needed to contain the core. Up to 120 workers were employed in the dam's construction. The boulders were lifted with levers on mining carts and then transported to the dam site.

Among the mining dams Ripking had seen in Sweden were possibly those near Sala, 110 km northwest of Stockholm (Engelbertsson 1991). There several reservoirs were impounded between 1504 and 1543 as far as 7 km northwest of the mine. The largest was the Sala lake some 4 km to the northeast, near the ore-treatment-plant which for its operation and the washing process required much more water than the pumps and lifting devices in the mine. The latter's reservoir system was expanded from 1595 to 1660 and again in 1819-1822. In the peak period there were 62 lakes covering 56 km^2 of which many have been drained since then. Also many of the canals feeding or interconnecting them fell into disuse.

A fourth mining district that made ample use of water power was the one of Banská-Stiavnica, some 130 km east of Bratislava in Slovakia (Novak 1972). The earliest reservoirs were used since 1510 either as fish ponds (see below) or for water supply. Water power was introduced towards the end of the 16th century and the first power embankments on which firm data are available date from 1614 (Fig. 99, Table 21). At the beginning of the 18th century, mining was interrupted due to dewatering problems. They were solved by Mateja K. Hell (1653-1742), chief mechanic. Moreover, he had the Eviçka dam heightened and some new ones built. The greatest dam building activity however occurred under the direction of Samuel Mikovini (1700-1750), responsible for the

Swedish mining dams

Power dams in Slovakia

Figure 99. Map of the region of Banská-Stiavnica in southern Slovakia showing the mining reservoirs and the canals (after Peter & Lukáč 1972).

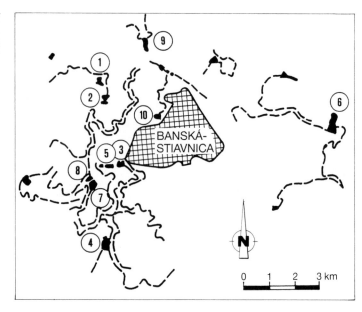

Table 21. Important power embankments around Banská-Stiavnica in southern Slovakia.

No. in Fig. 99	Year of completion	Name of dam	Height (m)	Length (m)	Reservoir capacity (million m^3)
1	1614	Dolná Hodruša	22	199	0.64
2	1614/1745	Horná Hodruša	16/22	185	0.25
3	1638/1714	Evicka	11	165	0.21
4	1738/1779	Pocúvadlo	11/23	183/215[1]	0.75/0.92[1]
5	1739	Bakomi	16	114	0.18
6	1740	Bansky Studenec	17	167	0.75
7	1740	Large Richnava	23	180	0.77
8	1740	Small Richnava	17	188	0.55
9	1744	Rozgrund	30[1]	124	0.49
10	Bef. 1770	Klinger	22	127	0.16

[1]Record for this group of dams.

Rozgrund dam

embankments Nos. 4, 5 and 7 to 9 in Table 21. Among these, the most remarkable was the 30 m high Rozgrund dam north of Banská-Stiavnica, charateristic for its steep outer slopes and thin clay core (Fig. 100). Its height had already been surpassed some 70 years earlier by the French navigation dam St. Ferréol (see below), but in boldness it was quite unique. The structure is still in operation today, for the purpose of water supply (ICOLD 1973, 1984).

A number of unusual mining dams were built at the same time approximately 70 km southeast of Timşoara in southwestern Romania, after this area had (like Slovakia) become part of the Austro-Hungarian empire (Table 22) (Botzan et al. 1991). Details are

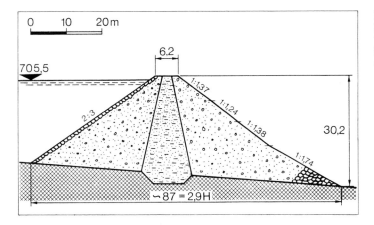

Figure 100. Cross section of the Rozgrund dam completed in 1744 north of Banská-Štiavnica in southern Slovakia (after Peter & Lukáč 1972).

Table 22. Masonry power dams in southwestern Romania.

Year of completion	Name of dam	Type	Height (m)	Length (m)	Reservoir capacity (million m³)
1733	Large Oravita	Buttress	13	117	0.14
1733	Small Oravita	Gravity	9	67	0.05
Ca. 1740	Large Dognecea	Buttress	14	71	0.55
Ca. 1740	Small Dognecea	Gravity	12	57	0.15

available for the two Dognecea dams. While the smaller one was a rather slim gravity dam with an earth core and an embankment added on the upstream (!) side, the large Dognecea dam had a pure gravity section with a crest width of 6 m and an inclination on both faces of 35%. However, the plans were the most remarkable aspect of the dams, whereof especially the larger one exhibited two downstream buttresses, between which the upstream face was curved, while the downstream one was straight (Fig. 101).

Mining dams in Romania

In the wake of Peter the Great's (1672-1725) modernization of Russia, several dams were built between the Ural and the Altay Mountains in some of the mining districts of Siberia (Table 23) (Danilevskii 1968). The emperor himself hired the young Dutchman Georg W. Hennin (ca. 1680-1750). He designed the first dam in Yekaterinburg in the centre of the Ural Mountains. It consisted of two rows of double timber cribs. Together with the intermediate space they were filled with clay. The dam's base width attained 43 m or over six times its height. In the middle there was a gated spillway chute and on either side outlets to two canals that, in turn, fed a total of 50 water wheels. The Yekaterinburg layout was also adopted at most of the other Siberian dams, but they consisted usually of homogeneous embankments. The Zmeinogorsk dam built in the 1780's at the western foot of the Altay Mountains was the largest. It required about 100 000 m³ of fill (Fig. 102).

Russian power dams

Figure 101. Downstream view of the large Dognecca dam in southwestern Romania (photo A. Vogel, Vienna).

Table 23. Russian power dams in western Siberia.

Year of completion	Name of dam	Distance in km from city	Type	Height (m)	Length (m)
1723	Yekaterinburg		Crib	6.5	209
1729	Kolyvan	420 S Novosibirsk	Embankment	7.0	110
Ca. 1760	Votkinsk	400 W Yekaterinburg	Embankment	10.0	685
1761	Zlatoust	200 S Yekaterinburg	Embankment	9.2	266
Ca. 1780	Zmeinogorsk	430 S Novosibirsk	Embankment	18.0	158
1793	Omsk	800 E Yekaterinburg	Crib	8.5	140

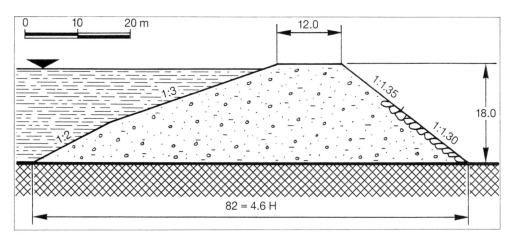

Figure 102. Cross section of the Zmeinogorsk dam at the western foot of the Altay Mountains in Russia (after Danilevskii 1968).

5.1.3 *Other postmedieval power dams*

After the Middle Ages, dams also serving as mill ponds continued to be built throughout Europe. The few on which sufficient data are available are shown in Table 24 (Botzan et al. 1991, Breznik 1984, Fernández et al. 1984, García-Diego 1971, 1977, Quintela

Table 24. Other important European power dams (from 1500 to 1815 of which data are known).

Year of com-pletion	Name of dam	Distance in km from city/country	Type	Height (m)	Length (m)	Reservoir capacity (million m^3)
1570	Casillas II	210 N Sevilla/Spain	Buttress	9	80	0.10
1596	Kobila[1]	36 W Ljubljana/ Slovenia	Crib	10	20	?
1610	Eastern Kreuz	1 E St. Gall/Switzerland	Embankment	7	60	0.06
1615	Knaben	1 E St. Gall/Switzerland	Embankment	7	70	0.05
1690	San Jorge	230 N Sevilla/Spain	Buttress	11	175	0.30
1693	Casabaya	122 NW Sevilla/Spain	Buttress	16	120	0.10
1713	Mannen	1 E St. Gall/Switzerland	Embankment	9	70	0.06
1735	Bedia	12 SE Bilbao/Spain	Buttress	5	56	small
1747/1950	Feria	135 NW Sevilla/Spain	Buttress	19/24	149	0.95
1749	Monte Branco	145 E Lisbon/Portugal	Gravity	12	?	?
After 1750	Generala	220 N Sevilla/Spain	Gravity	10	200	0.20
Before 1780	Penedos	100 E Lisbon/Portugal	Gravity	13	120	?
1788	Large Taul[1]	310 NW Bucharest/Romania	Embankment	28	172	0.16

[1]Mining power dam.

et al. 1988, Schnitter 1992, Votruba 1987). The eastern Kreuz dam and its western counterpart as well as two adjoining embankments built in 1615 and 1713 respectively provided water not only for mills, but also for wetting the linen during the bleaching process in the sunlight, an important operation in the textile industry of St. Gall in Switzerland (Schnitter 1992).

Bleaching ponds

The dams in southwestern Spain and Portugal continued the progress of mill dams that was iniciated by the medieval Castellar dam (see above) (Fernández et al. 1984, García-Diego 1977). Like the latter, the Casabaya dam had to rely on the end and intermediate walls of the mill house at its toe for stability. But already from 1565 to 1570 the young architect Francisco Becerra (1545-1605) built the Casillas II dam 28 km southwest of his native town Trujillo in the old Roman manner, i.e. a rectangular water retaining wall propped up by rather closely spaced buttresses (Fig. 103). In 1572 he started the construction of the larger San Jorge dam 2 km southwest of Trujillo, but he left for the Spanish colonies in America, where he became a famous architect. The dam was only completed in 1690 and it had such a sturdy water retaining wall, that the widely spaced buttresses were actually unnecessary. This also held true for the Feria dam originally some 19 m high (Fig. 104). The spaces between five of its seven massive buttresses not only contained two mills but, since the dam's construction was ordered by a bishop, they were also enhanced by a chapel!

Revival of buttress dams in Spain

San Jorge and Feria dams

A few years prior to the construction of the Feria dam, five smaller structures were built near Bilbao in northwestern Spain. For the first time since the Roman Esparragalejo dam in south-

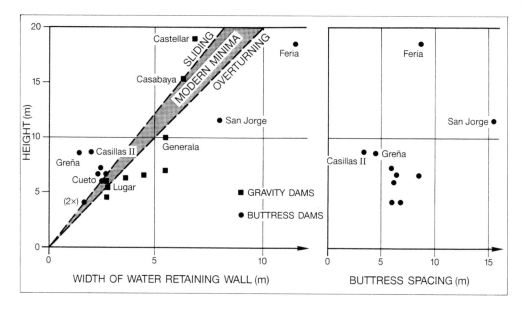

Figure 103. Width and buttress spacing of mill dams in south-western Spain in function of their height.

Figure 104. Downstream view of the Feria buttress dam in southwestern Spain; the spill-ways at both ends were added on occasion of a later heightening (photo by the author).

Multiple arch dams

western Spain (see above) they revived the multiple-arch buttress type (Fernández et al. 1984, García-Diego 1971-1972). Actually, this concept had already been proposed two hundred years earlier in 1530 by the architect Baldassare Peruzzi (1481-1536) in Siena/ Italy for the reconstruction of the failed fish pond on the river Bruna (see below) (Adams 1984). He envisaged a series of arches supporting a slab, which sloped upstream (top of Fig. 105). Around thirty years later, the idea reappeared in 'The Twentyone Books of Devices and of Machines', an unpublished Spanish codex traditionally (but falsely) attributed to the Italo-Spanish hydrauli-

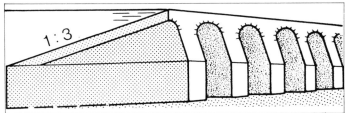

Figure 105. Multiple-arch dams proposed in 1530 by Baldassare Peruzzi (top, from Adams 1984) and around 1580 in 'The Twentyone Books of Devices and of Machines' (bottom, after García-Diego 1976).

cian Juanelo Turriano (1511-1585) (García-Diego 1976). Their author correctly noted that the water load on the sloping upstream face would increase the dam's stability (bottom of Fig. 105).

This recommendation was followed by the builder of the Bilbao dams, the Basque nobleman Pedro B. Villareal de Berriz (1670-1740). He was interested in water power and published a relevant book in 1736 (Villareal 1973). However, the downstream faces of his arches were vertical like at Esparragalejo, so that their width increased considerably between the dam's crest and base and even more so towards the buttresses (Fig. 106). All five structures were less that 5 m high and consisted of one to five arches with spans between 15 m (one arch) and 12 m (five arches). They acted as diversion weirs to headrace canals and in view of their sturdy construction all excess flow was discharged over the crest. With the exception of one which suffered from the chemical action of waste-waters, they all are in an excellent state of preservation.

Villareal's dams

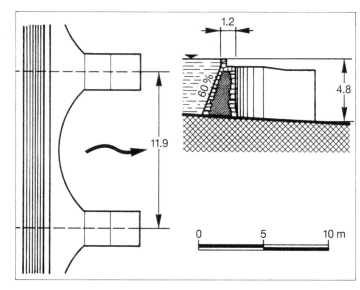

Figure 106. Plan of one of five spans and cross section of the Bedia multiple-arch dam near Bilbao in northwestern Spain (after Fernández et al. 1984).

5.2 FISH PONDS

Rise of pisciculture

Several of the abovementioned medieval power dams were concurrently used for pisciculture, especially when they belonged like the Lucelle dam in northwestern Switzerland (Table 19) to a Cistercian monastery, where partaking of meat was forbidden (Schnitter 1992). But also the nobility enjoyed eating fish no less than the surviving citizens who had become rich by inheritance after the Great Plague of 1347-1350. It had meanwhile become economically interesting to transform fields and meadows into fish ponds, a process that required considerable initial investments, however, and these means were usually available only to monasteries, cities and the noblesse. Moreover, fish ponds provided a sure and continuous source of food, particularly in times of war. In the late Middle Ages, and up to about the 16th century, fish ponds were built in many parts of Europe. Those, for which sufficient data are available, are listed in Table 25 (Adams 1984, Binnie 1987, Botzan et al. 1973, ICOLD 1973, 1984, Lanser 1962, Poláček & Dvořáčková 1991, Schmidt & Hobst 1991, Schneider 1989, Schnitter 1992, Votruba 1987). All these reservoirs were im-

Bruna dam

pounded by homogeneous embankments, with the exception of the important Bruna gravity dam built by the Italian city of Siena and which failed on its first impoundment as already mentioned (Adams 1984). In spite of several proposals to do so, it was never reconstructed.

The greatest number of fish ponds were constructed in Bohemia, some of which are of considerable size (Kratochvil 1967,

Table 25. European fish ponds (of which data are known).

Year of completion	Name of pond or dam	Distance in km from city/country	Type	Height (m)	Length (m)	Reservoir capacity (million m^3)
Ca. 1180	Kilburn	90 S Newcastle/GB	Embankment	7	55	?
Ca. 1200	Alresford	87 SW London/GB	Embankment	6	70	2.4
Ca. 1230	Oldstead	90 S Newcastle/GB	Embankment	8	400	?
1272	Máchovo	60 N Prague/Czech.	Embankment	9	?	5.5
Before 1367	Dvořisté	116 S Prague/Czech.	Embankment	10	520	6.7
End 14th cent.	Nucet	70 NW Bucharest/Romania	Embankment	8	?	Ca. 5.0
1430	Hauptwil[1]	12 NW St. Gall/Switzerland	Embankment	4	80	0.1
1460	Spiegelfreud	48 W Innsbruck/Austria	Embankment	8	250	?
1470	Upper Nabern	30 SE Stuttgart/Germany	Embankment	6	100	?
1492	Bruna[2]	93 S Florence/Italy	Gravity	18	300	?
1512	Horusický	103 SE Prague/Czech.	Embankment	11	714	4.0
Before 1549	Staňkovský	130 SE Prague/Czech.	Embankment	15	170	20.0
1574	Svět	124 S Prague/Czech.	Embankment	9	1575	?
1587	Dorohoi	395 N Bucharest/Romania	Embankment	10	?	Ca. 6.0
1590	Rožmberk	118 S Prague/Czech.	Embankment	11	2430	13.8
16th cent.	Dehtár	120 S Prague/Czech.	Embankment	10	230	4.8
16th cent.	Mutiněveský	112 SE Prague/Czech.	Embankment	11	310	1.5

[1]Lowest of five similar ponds in series. [2]Failed on first impoundment.

Votruba 1987). In the 15th and 16th centuries some 700 such ponds were built, which covered a total area of 1800 km^2 or over three times the surface of Lake Geneva! The reservoirs were often interconnected into veritable systems by common feeder canals, such as shown in Figure 107 for the ponds around Třeboň, 120 km south of Prague. There, the 43 km long Golden Canal was built under the direction of Josef N. Štěpánek (1505-1538). His successor Jakub Krčín (1535-1604) was not only responsible for the 2430 m long Rožmberk dam, but he also constructed the 11 km long diversion canal from the Lužnice river to the Nežárka with the purpose of keeping floods away from his new pond. Similar systems of feeder canals and ponds were built around Poděbrady and Pardubice, 50 and 100 km east of Prague respectively.

As mentioned earlier, all these Bohemian embankments were homogeneous and consisted of sandy clay spread out in layers of 0.15 to 0.20 m height and compacted with wooden tampers (Schmidt & Hobst 1991). While the upstream slopes were rather steep and protected by a layer of dry masonry (Fig. 108), the downstream ones were often flattened by means of intermediate horizontal berms. Overflow spillways were usually arranged at one end of the dams and provided with racks in front of them so as to prevent the escape of fish. These racks were triangular or pentagonal in plan. At one of the two spillways of the Rožmberk dam, the

Bohemian fish ponds

Rožmberk dam

Embankment construction

Figure 107. Map of the region of Třeboň south of Prague, showing the fish ponds and the feeder and diversion canals (broken lines) (after Votruba 1987).

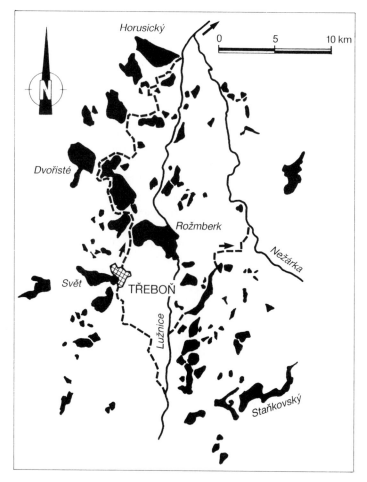

Figure 108. Part of the upstream face and outlet structure of the Rožemberk dam in Bohemia (photo by the author).

2.5 m high wooden rack was 234 m long! The outlets through the dam bases consisted of two trunks. Channels of 0.20 to 0.40 m width and 0.18 to 0.35 m depth had been cut into them and they were tied together to form a sort of 'pipe'. Closure was provided by small spoon-like gates slid along the dams' upstream faces by means of poles. The Rožmberk dam had not less than 20 such outlet 'pipes'!

In 1547, under the pseudonym Dubravius bishop Jan Skála (1480-1553) published 'Five Books on Fish Ponds' in Latin. They were, however, soon printed in English, German and Polish as well. His works synthesized the Bohemian experience and knowledge in pisciculture and the construction of fish ponds. The latter came to an end during the 30 years-war 1618-1648 and, thereafter, many ponds were abandoned or even drained to reclaim farm land.

In 1600 the Surveyor General of the King's Woods south of the river Trent in central England, John Taverner (died 1606), published his 'Certaine Experiments concerning Fish and Fruite' (Binnie 1987). In his book, he also set rules for the design and construction of embankment dams, stipulating a crest width equal to their height and a three times wider base. These were the very dimensions applied for the mining dams in the German Harz Mountains. Likewise, the type of bottom outlet and their closing device recommended by Taverner were very similar (see above). Moreover, his design recommendations for the dams were identical to those of Dubravius, whose book on fish ponds had also been published in English approximately fifty years earlier, as we have seen (Votruba 1991).

Outlet works

Earliest books on dam engineering

5.3 IRRIGATION DAMS

Although irrigation was practiced in many parts of medieval and postmedieval Europe to moisten and fertilize meadows, it was essential and required storage reservoirs in the southern countries like Italy and Spain. The data available on dams built in this respect are summarized in Table 26 (Fernández et al. 1984, ICOLD 1973, 1984, Smith 1971). The dams for power and pisciculture in southern countries described earlier, as well as those built for irrigation purposes were mostly masonry structures, and this in contrast to the embankments prevailing in northern Europe. Possibly the preference for masonry dams may be seen as an inheritance from the Romans who used embankments almost exclusively as supporting elements. In Spain this 'tradition' was continued in the Moslem weirs (see above).

Need for irrigation reservoirs

Masonry versus embankment dams

Table 26. European irrigation dams.

Year of completion	Name of dam	Distance in km from city/country	Type	Height (m)	Length (m)	Reservoir capacity (million m^3)
13th/18th cent.	Almonacid	40 S Zaragoza/Spain	Gravity	25/30	104	silted up
1384/1586	Almansa	94 SW Valencia/Spain	Gravity[1]	17/23	40/89	?/2.8
1560	Granjilla 2	40 NW Madrid/Spain	Gravity[2]	6	460	0.2
1572	Ontigola	45 S Madrid/Spain	Buttress	12	280	?
1594	Tibi	110 S Valencia/Spain	Gravity[1]	46	65	5.4
Ca. 1600	Ternavasso	29 SE Turin/Italy	Embankment	7	326	0.3
1640	Elche	130 S Valencia/Spain	Arch	23	95	0.4
1660	Granjilla 1	40 NW Madrid/Spain	Gravity[2]	14	239	0.4
1704/1929	Arguis	82 N Zaragoza/Spain	Gravity	23/28	35/37	1.1/2.7
1766/1841	Caromb (Paty)	90 N Marseilles/France	Gravity	17/22	65	1.2
1776/1879	Relleu	100 S Valencia/Spain	Gravity[1]	29/33	34	0.6
1791	Puentes	67 SW Murcia/Spain	Gravity	50[3]	283	52.0
1806	Valdeinfierno	74 W Murcia/Spain	Gravity	30[4]	101	4.0

[1]Curved in plan. [2]With supporting embankment. [3]Failed 1802 and rebuilt 1884 to 69 m height. [4]Heightened 1897 and 1965 to 49 m total height.

5.3.1 *Revival of the arch dam*

Moslem influence in Spanish irrigation

Moslem irrigation techniques were maintained in Spain even after the progressive Christian reconquest (1031-1492) (Glick 1970). Therefore, the abovementioned upsurge of dam construction in the 13th/14th centuries under the Mongolian Ilkhans in Iran – which at the time was prominent in the Moslem world – conceivably had its repercussions also in Spain. In fact, the first large storage dam for irrigation, Almonacid de la Cuba, was built south of Zaragoza in northeastern Spain under the Aragonese king Jacob I, the Conqueror (1213-1276). It was a 25 m high gravity dam of rectangular cross section with a straight alignment (Fernández et al. 1984).

But the Almansa dam completed in 1384 in southeastern Spain to a height of 17 m was already curved in plan, like the abovementioned Ilkhanid arch dams (Fig. 109). Its cross section was, however, considerably sturdier, since the crest width already amounted to some 9 m with a stepped downstream face whereby the base width reached some 16 m (Fig. 110). The radius of curvature at the crest was about 31 m and the central angle almost 100°. In 1586, the dam was heightened some 6 m by building a polygonal gravity wall of nearly rectangular cross section on the crest. For its construction, material was taken from the upstream side of the old dam where it protruded beyond the upstream face of the heightening. In the central part, the arch was thus reduced to less than half of its

Almansa curved gravity dam

Figure 109. Downstream view of the Almansa dam in southeastern Spain showing the arch of 1384 and the polygonal gravity dam added in 1586; the upstream gate tower was built in 1911 (photo by the author).

Figure 110. Cross sections of four arch dams in southeastern Spain (after Fernández et al. 1984).

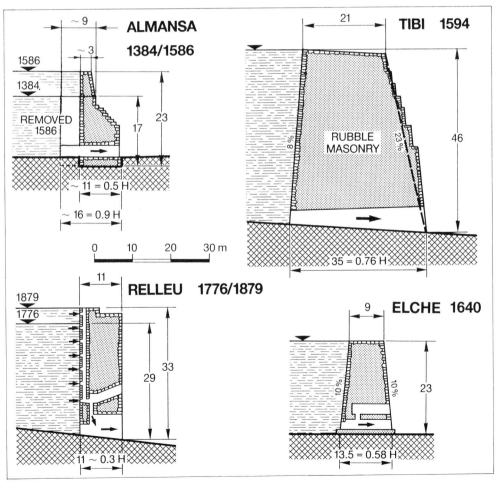

original width, but the remainder was still amply sufficient in view of the strong curvature.

A novel feature at Almansa, later adopted in all Spanish masonry dams, was the large bottom outlet of over 3 m² clear opening for purging sediment deposits from the reservoir. It was closed near the upstream inlet by propped up boards, the removal of which was very dangerous even at a drawdown reservoir. Very often this work was therefore assigned to people condemned to death and who were granted a pardon, if they survived the operation at all. Spilling of floods was originally over the dam's crest and a spillway chute at its right end was added only during the heightening of 1586.

The latter fell into the reign of Philip II (1556-1598) when Spain attained the peak of its power and wealth. Philip either initiated or supported the construction of several dams, among them the enormous Tibi dam in southeastern Spain. Upon request by the local population, he had his clockmaker and hydraulic expert Juanelo Turriano, mentioned earlier, check the design and construction started in 1580. It was completed as late as in 1594, by when the structure had reached 46 m of height and a volume of over 36 000 m³ of rubble masonry and ashlar lining. It had a very sturdy cross section (Fig. 110), in spite of it being curved in plan to a crest radius of 97 m with a central angle of 55°. Similarly to the Iranian dams and other precursors, water was drawn from the reservoir by means of a shaft near the upstream face with 51 (!) small inlets at different levels. As at Almansa, there was an ample bottom outlet that widened on its downstream side to facilitate the aeration of the outflow. For that purpose a separate gallery was provided above the bottom outlet in later dams, thus also improving the safety of the opening operation (Fig. 110).

About half a century after Tibi, the first true arch dam in Europe since Roman times (see above) was built from 1632 to 1640 near Elche, farther southwest. It was designed by one Joanes del Temple who adapted it masterfully to the peculiarities of the site with one flank consisting of a low ridge (Fig. 111). He let the arch across the main gorge end in an artificial abutment block on the ridge, from where a long wing wall completed the structure. Also, the other end of the arch crest abutted not against rock, but against a wing wall directed upstream.

Thus, in the Elche dam's uppermost part, the arch thrusts did not have much support and were therefore deflected downwards. This had been the case to an even greater extent at the Mongolian Kebar dam and the Monte Novo dam believed to be of Roman origin (see above). The concept was also shown in an unpublished patent accorded in 1606 to Jerónimo de Ayanz (ca. 1500- after 1610) by the Spanish king (García 1990). Elche's main arch was

Large bottom outlet

Huge Tibi dam

Elche true arch dam

Masterful design

Figure 111. Upstream view of the Elche true arch dam in southeastern Spain (photo by the author).

some 75 m long at the crest, which was curved to a radius of 62 m with a central angle of 70°. The dam's rather slim cross section (Fig. 110) was completely adequate as verified with a computer analysis by the crown–cantilever method: no vertical tensile stress at the heel and a maximum compression of 0.8 MPa at the toe; half as much horizontal compression and up to 0.2 MPa tension in the relatively clumsy arches (without thermal effects).

In contrast to this elegant structure, the Relleu dam, begun possibly as early as 1653 but completed only in 1776, was a crude affair and marked the end of pre-modern arch dam construction in southeastern Spain. While its vertical downstream face was slightly curved, the upstream one formed a vertical plane. Therefore, the stability of the rather thin dam with a length-to-height ratio of only 1.0 relied not on arch action, but rather on that of a wedged-in triangular plate (Fig. 110).

Crude Relleu dam

5.3.2 *Other postmedieval irrigation dams*

Under the rule of the Spanish king Philip II, another Roman dam building concept was revived, i.e. the embankment supported water retaining wall (see above). Near his Escorial palace, 40 km northwest of Madrid, the 7 m high Granjilla 2 dam was erected in 1560 for the irrigation of the royal gardens (Fig. 112) (Fernández et al. 1984). Its reservoir was also used for recreation and aquatic festivals, as witnessed by a small square island built in front of the dam and connected to its crest by a 50 m long and 2 m wide stone bridge. One century later, a similar dam, Granjilla 1, was constructed not far upstream. It was twice as high as the former and the supporting embankment consisted of rockfill instead of the usual earthfill (Fig. 112).

Revival of Roman embankment type...

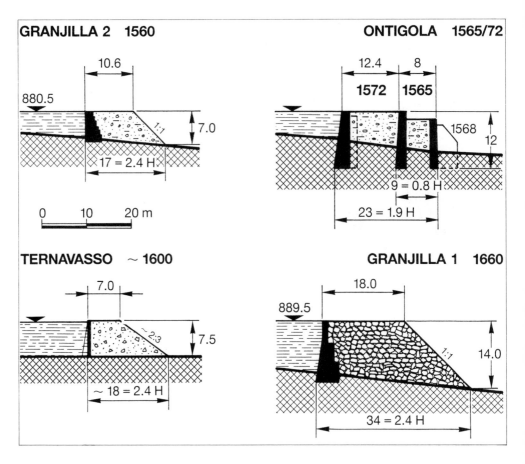

Figure 112. Cross sections of four embankment dams in central Spain and northern Italy (after Díaz-Marta 1992, Fernández et al. 1984 and Smith 1971).

...and ancient gravity dam with earth core

A third ancient concept, that of the gravity structure with an earth core, later taken up at the Potosí/Bolivia and the abovementioned Oder mining dams, was revived at Ontigola (Díaz-Marta 1992, García & Rivera 1985). This dam was completed in 1565 to the design and under the direction of Juan B. de Toledo (ca. 1515-1567) for the irrigation of the royal gardens at Aranjuez, 45 km south of Madrid. Although Toledo had gathered some experience in hydraulic works in southern Italy dominated by Spain at the time, he sought the advice of dikemasters from the also Spanished ruled Netherlands. The result was an unusually slim structure, the downstream wall of which started to tilt as soon as the dam was impounded (Fig. 112). As a remedy, some downstream buttresses were added until 1568, when a large section of the upstream wall failed. It was replaced under the direction of Juan de Herrera (1530-1597) by a new wall and core in front of the old one, which brought the total width of the structure to a comfortable 190% of its height.

About 30 years later, a dam of the embankment supported wall type was completed near Turin in northern Italy (Smith 1971). It had the distinction that its merely 0.6 m thick water retaining wall was built of bricks and, on the upstream face was provided with buttresses to keep it from falling into the reservoir upon the latter's drawdown, very much like the Roman Proserpina dam (see above) (Fig. 112).

The 18th century also saw the construction of some important gravity dams. The construction of the Arguis dam for irrigation around the city of Huesca, 330 km northeast of Madrid, took from 1687 to 1704 (Fernández et al. 1984). It was designed by a professor of mathematics at that city's university, one Francisco A. de Artigas. Its rectangular cross section was changed in modern times to a 5 m higher triangular one (Fig. 113). Also, the Caromb (or Paty) irrigation dam built much faster between 1764 and 1766, was designed by a mathematician, Jean-Claude Morand (1707-1780), teacher at the Jesuit college in nearby Avignon in southeastern France

Important pure gravity dams

Figure 113. Cross sections of three 18th century gravity dams in France and Spain (after Caromb 1989, Fernández et al. 1984, Martin & Muñoz 1986).

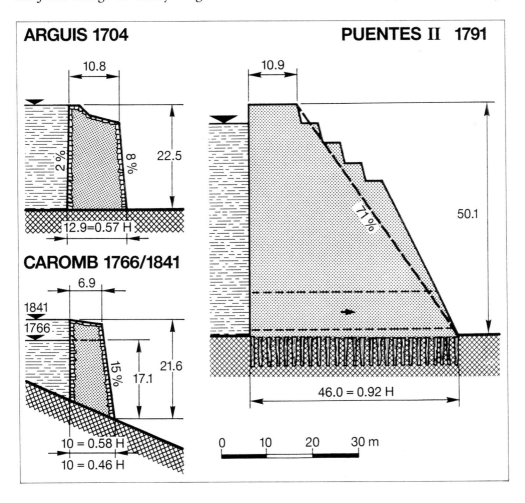

Figure 114. Downstream view of the Caromb (or Paty) gravity dam in southeastern France after its reinforcement in 1977 (photo by the author).

(Caromb 1989). Its precarious stability was aggravated by a heightening in 1841 and remedied only in 1977 by means of a downstream reinforcement (Figs. 113, 114).

Puentes tragedy

The most ambitious, and also one of the most unfortunate projects of the period, was unquestionably the Puentes irrigation dam in southeastern Spain (Martin & Muñoz 1986). The foundation was laid for a 26 m high gravity wall during winter 1647-1648. It was washed away by a flood in the following summer. This could happen only because it bridged over the alluvial deposits in the river which were too deep to excavate with the means available at the time. Incredibly enough, the difficult site was retained, when about 140 years later construction was started in 1785 on a second Puentes (= bridges) and the nearby Valdeinfierno (= valley of hell) dams. The designer of the grandiose scheme, Gerónimo Martínez de Lara (1750-1814), a native of Murcia, had the new, massive Puentes dam of some 50 m of height founded on wooden piles driven into the alluvial deposits (Fig. 113).

Difficult foundation

When the reservoir's water level reached 93% of the dam's height for the first time early in 1802, a sudden washout of the alluvial deposits occurred. An 18 m long and 34 m high hole was torn in the lower part of the dam releasing 30 million m^3 of water within one hour (Fig. 115). In the city of Lorca 608 people drowned in the ensuing flood and a further 328 in the surroundings. In 1881-1884, construction of the Puentes dam was resumed for the third time, some 200 m downstream of the original site. Although the valley was wider, there was a rock outcrop in the river bed at this location. The Puentes failure, an innocuous one at Gasco (see below) and the rapid silting up of the Valdeinfierno reservoir marked a low point of dam engineering in Spain, anticipating the country's misery during the Napoleonic wars (1808-1813).

Washout causes catastrophic flood

Figure 115. Downstream view of the Puentes dam in south-eastern Spain some 50 years after its failure (from Martin & Muñoz 1986).

5.4 WATER SUPPLY DAMS

The Roman standards of public supply with running fresh water were again attained in Europe only from the 13th century onwards, when the first cities began to build (mostly wooden) pipelines to harness springs within or from without their walls (Grewe 1991). Since only a fraction of the discharge was usually diverted, reservoirs were rarely needed before the 19th century. The important ones, for which sufficient data are available are listed in Table 27 (Binnie 1987, CDC 1967, Çeçen 1987, ICOLD 1973, 1984, Öziş 1977, Votruba 1991). Europe's oldest water supply dam since Roman times appears to be Jordán completed in 1492 near the Southbohemian city of Tabor, centre of the Hussite religious revolution (1419-1485) (Votruba 1991). The embankment had steep 1:1.8 and 1:1.6 up- and downstream slopes and was built in the same way as the abovementioned Bohemian fish pond dams. To reach the city, the water had to be pumped up 32 m. The water wheel driving the three pistons of the pump consumed some of the water supplied in the process.

Most of the other early water supply dams were built for Istanbul in Turkey, the capital city of the eastern Roman and, later, the medieval Byzantine empires. In those times the compensation between supply and demand was accomplished by numerous open and underground cisterns within the city walls (Forchheimer & Strzygowski 1893). After the conquest of Istanbul in 1453, the Ottoman Turks reconstructed some of the old aqueducts that brought water into the city and built new ones (Çeçen 1987). The famous architect Sinan (1495-1588) restored the 22 km long, sixth century Kirkçeşme conduit from the north and prolongated it some 20 km into the Belgrade forest between 1554 and 1560 (Fig. 116). Here, a first storage dam was built in 1560 on the remains of

Revival of public water supply

Jordán dam

Istanbul supply system

Belgrade dam

Table 27. European water supply dams.

Year of completion	Name of dam	Distance in km from city/country	Type	Height (m)	Length (m)	Reservoir capacity (million m^3)
1492	Jordán	75 S Prague/Czech.	Embankment	18	300	3.0
1560[1]	Belgrade (Büyük)	21 N Istanbul/Turkey	Gravity	15	85	1.3
1620	Topuz (Karanlik)	23 N Istanbul/Turkey	Gravity	10	65	0.1
1750/1786	Topuzlu	23 N Istanbul/Turkey	Buttress	12/16	81	0.2
1751	Pilsko	67 SW Prague/Czech.	Embankment	16	396	0.7
1765	Ayvat	24 N Istanbul/Turkey	Gravity	15	66	0.2
1796	Valide	24 N Istanbul/Turkey	Buttress	14	104	0.3
1810	Hámori-Gát	145 NE Budapest/Hungary	Embankment	18	100	0.5

[1]Failed and rebuilt 1748.

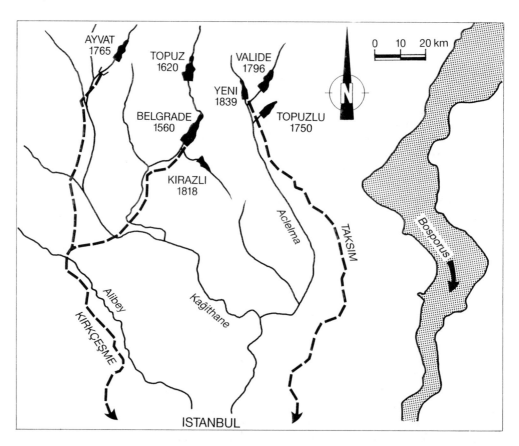

Figure 116. Map of the water supply reservoirs and conduits (broken lines) in the Belgrade forest north of Istanbul in Turkey (after Öziş 1977). a Roman forerunner and reconstructed in 1724, 1748 and 1900 (Çeçen 1987, Öziş 1977). It is now a fairly massive gravity dam with a height of 15 m and a base width of 65% of the height. The similar, though smaller Topuz (or Karanlik) structure of 1620, was provided with some rather unnecessary downstream buttresses.

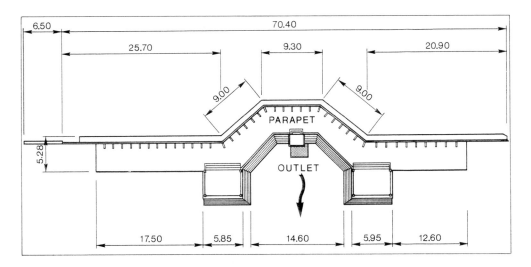

However, buttresses were needed at the Topuzlu dam built in 1750 for the new Taksim aqueduct commissioned in 1731. In fact, the almost rectangular wall had a base width of just 44% of the height after its heightening in 1786. Unusual, and reminiscent of the only about 10 years older Dognecea mining dams in south-western Romania (Turkish until 1718, see above) was the plan of the Topuzlu dam with an arch-like upstream bulging of the central part between the buttresses (Fig. 117). This same design was adopted for the Ayvat and Valide dams built within the next half century. The last two structures of the system – built in the early 19th century – had again rectangular gravity sections. The 17 m high Yeni (or Sultan Mahmut) dam completed in 1839 however, was curved in plan to a radius at the crest of 62 m with a central angle of 82°. Since its base width amounted to just 53% of the height, it relied to some extent on arch action for its stability.

The hydraulic features of all seven dams followed practically the same pattern over almost three centuries. Only for the spillways there were two different principles: either an overflow with chute at one or both ends of the dam or a conduit from one end, which had its intake between normal storage water level and the dam crest. In relation to the hydrologically similar catchment areas, the spillway discharges varied widely from 0.09 to 5.68 m³/s km² (Öziş 1977). Also, the diameter of the bottom outlet pipes of 0.15 to 0.25 m had little or no reference to either the dam height or its reservoir content (Çeçen 1987). The same was true for the parallel pipes of 0.10 to 0.25 m diameter for water withdrawal. In order to be able to operate their valves according to demand, they discharged into small measuring basins, with 6 to 18 orifices calibrated for flows between 52 and 1300 m³/day (Fig. 118).

The maximum withdrawals possible from the reservoirs varied

Figure 117. Plan of the Topuzlu buttress dam north of Istanbul in Turkey (after Çeçen 1987).

Six additional storage dams

Spillways

Outlets

Figure 118. Outlet of Yeni dam north of Istanbul in Turkey, showing the measuring basin with 11 orifices calibrated for discharges from 52 to 518 m³/day (photo by the author).

Reservoir operation

Later European water supply dams

Pools and fountains in parks

between 1240 and 13 000 m³/day and totaled 26 200 m³/day. At such rates, the reservoir contents would last for 36 up to 126, or, on average, for 87 days. Except for extremely dry periods, of course many additional springs fed the aqueducts directly. Of these Kirkçeşme delivered some 17 000 and Taksim 7000 m³/day to mosques, churches, synagogues, schools, government buildings and the like as well as to no less than 840 public fountains. After the completion of the Kirkçeşme conduit in 1560, every inhabitant of Istanbul thus already had roughly 0.25 m³/day of water at his disposal which amounts to 20% more than today's quantity (Çeçen 1987). However, one important difference must be borne in mind: that the water then ran continuously, whereas it's flow is now mostly controlled by waste reducing taps.

Few water supply dams were built in the rest of Europe before the end of the turmoil caused by the French Revolution and Napoleon I (1804-1815). Eventually the concentration of population in the course of the Industrial Revolution necessitated a rapid expansion of numerous water supply systems, particularly in Great Britain (Binnie 1987). Subsequently, the same became true in other coutries as well.

5.5 DAMS FOR PARKS

At various times peculiar kinds of water supply systems were developed for the pools and fountains of the palace gardens of ostentatious rulers. We have already mentioned the Ajilah dam at Nineveh/Iraq in connection with Sennacherib's artificial swamp as well as the damming of the Aniene river for a pleasure lake near emperor Nero's villa at Subiaco/Italy (see above). The largest venture of its kind ever undoubtedly comprised the pools and fountains in the park of Versailles, 18 km southwest of Paris, executed under the 'sun king' Louis XIV (1661-1715) (Loriferne 1963). In 1668, immediately after completion of the first pumping plant at a reservoir existing near the palace, additional water had to be fetched from the Bièvre river (Fig. 119).

Versailles supply

Ten years later, three larger reservoirs were completed west of the palace and, in 1685, several more followed to the south and southwest, in conjunction with the enormous and famous Marly pumping plant on the Seine river. In total, over a dozen major reservoirs could store almost 8 million m³ of water. The rather low dams that impounded them consisted of a masonry wall supported by a downstream embankment, in other words: the Roman dam type revived over one century earlier for the irrigation of the Escorial palace gardens in Spain (see above). The rectangular water retaining walls of Versailles however, had the peculiarity of a central core of clay which, moreover, was connected to the clay lining of the reservoir bottom beneath the upstream half of the wall. The latter's stability obviously was rather precarious, especially in case of a rapid drawdown of the reservoir. The reservoirs' design was primarily owed to scientist-priest Jean Picard (1620-1682) and surveyor Philippe La Hire (1640-1718), both working under the general supervision of the king's famous minister Jean-Baptiste Colbert (1619-1683).

Reservoir system

Dam design

Although the pools and fountains of Versailles were imitated – on a much more modest scale, of course! – for the palace gardens of several (mainly German) rulers, none of these installations ap-

Figure 119. Map of the water supply reservoirs and canals (broken lines) for the park of Versailles near Paris in France (after Buffet & Evrard 1950).

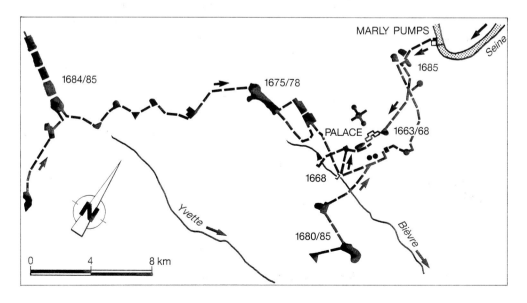

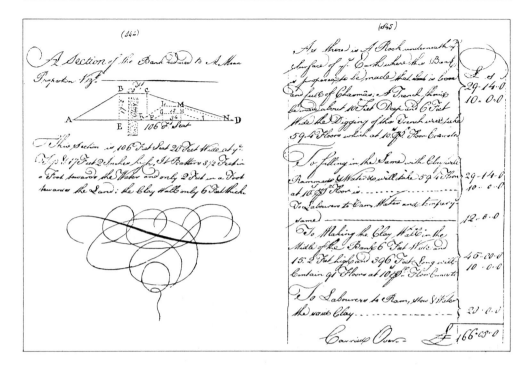

Figure 120. J. Grundy's cross section and cost estimate dated ca. 1749 for a proposed ornamental dam at Grimsthorpe in Great Britain (from Binnie 1987).

Ornamental dams in British parks

Reinvention of impervious core

pear to have included major reservoirs. In Great Britain artificial ornamental lakes became popular on many estates of the landowning class in the 18th century (Binnie 1987). The leading exponent of landscape gardening, Lancelot 'Capability' Brown (1716-1783), is said to have built over 40 such lakes, all of them impounded by earth dams of up to 7 m in height. For their impermeabilization he lined the upstream faces with 'puddle clay', i.e. clay thoroughly mixed with water and compacted with tampers. In the largest dam he built in the 1760's at Blenheim, 96 km northwest of London, Brown placed the puddle clay in the centre of the dam. However, John Grundy (1719-1783), hydraulic engineer, claimed to have adopted this technique for a small dam in Leicestershire already in 1741. He certainly did so at the 5.5 m high Grimsthorpe Great Water dam, 170 km north of London in 1748 (Fig. 120). Thus, the central impervious core introduced in 1558 at the mining embankments in the German Ore Mountains and later in the Harz Mountains and in Slovakia (see above) was reinvented in Great Britain, where it became a trademark of all major future embankments.

5.6 NAVIGATION DAMS

Already under the Romans, inland navigation played an important role in Europe. Inland waterways were the primary transportation method for bulky goods such as cereals, wine or construction materials. However, the small boats ordinarily could use the natural rivers. Therefore the Romans built only a few short navigable canals that served the access to harbours. These conditions prevailed until late in the Middle Ages, when the first important navigation canals were built in the Netherlands, northern France, Italy and Germany (Hadfield 1986). In the 17th century something like a 'canal fever' took hold of northern Europe and expanded as far south as Switzerland and southern France. Here, the 240 km long Languedoc canal (or Canal du Midi) connecting the Mediterranean Sea with the Atlantic Ocean, was the first to require a reservoir to compensate the losses of water incurred during locking operations in the dry season. Similar reservoirs followed on other canals. They are listed in Table 28 (Binnie 1987, ICOLD 1973, 1984, Smith 1971).

In fact , the proposal of a large water supply reservoir by a local landowner, Pierre-Paul de Riquet (1604-1680), turned out to be the key to the feasibility of an old intention, namely the one to build the Languedoc canal through a rather dry area (Rolt 1973). After approval of Riquet's plans by Colbert mentioned above and king Louis XIV, the engineer François Andréossy (1633-1688) prepared the designs for the canal and the reservoir in 1662. The construction got underway in 1666 under the supervision of

Importance of inland navigation

'Canal fever'

St. Ferréol water supply reservoir

Table 28. European navigation dams (1675 to 1815).

Year of completion	Name of dam	Distance in km from city/country	Type	Height (m)	Length (m)	Reservoir capacity (million m³)
1675	St. Ferréol	50 SE Toulouse/France	Embankment	36	780	6.7
1782	Lampy	61 SE Toulouse/France	Gravity	19	130	1.7
1797	Blackbrook[1]	155 NW London/GB	Embankment	11	?	0.6
1797	Rudyard	40 S Manchester/GB	Embankment	11	170	3.5
1797	Slaithwaite	30 NE Mancheste/GB	Embankment	21/22	140	0.3
1798	Tunnel End[2]	24 NE Manchester/GB	Embankment	13	?	0.1
1799	Diggle[3]	20 NE Manchester/GB	Embankment	14	?	0.1
1799	Gasco	25 NW Madrid/Spain	Buttress	87[4]	230	?
1799	Upper Chelburn	22 NE Manchester/GB	Embankment	17	165	0.1
1800	Hollingworth	19 NE Manchester/GB	Embankment	11	500	2.6
1803	Blackstone	24 NE Manchester/GB	Embankment	13	510	1.0
1805	Combs	30 SE Manchester/GB	Embankment	16	304	1.5
1810	Redbrook	21 NE Manchester/GB	Embankment	12	?	0.3
1811	Couzon	32 SW Lyon/France	Embankment	35	210	1.5

[1]Breached by floods 1799 and 1801; rebuilt as gravity dam 1906. [2]Damaged by flood 1799; rebuilt 1806. [3]Damaged by floods 1799 and 1806; replaced about 1830 by new dam upstream. [4]Construction discontinued 1799 at 56 m height; never in use.

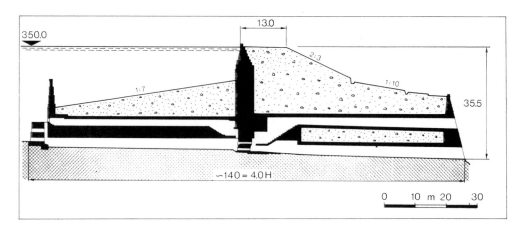

Figure 121. Cross section of the St. Ferréol navigation dam in southwestern France (after Ziegler 1900-1925).

Riquet. Like the Versailles dams that were to follow, the huge St. Ferréol dam consisted of a water retaining wall supported by a downstream embankment (Fig. 121). To stabilize the water retaining wall in case of too rapid a drawdown of the reservoir, a lower embankment was built against its upstream face. The result may also be regarded as an earth dam with a masonry core. With its height of 36 m, the St. Ferréol dam remained the highest embankment in the world for 165 years!

Couzon and Lampy dams

About a century later, the St. Ferréol structure was duplicated with the Couzon canal dam in southeastern France whose height was inferior by only 1 m (Smith 1971). Whereas its construction began in 1788, it reached completion as late as 1811. At the same time an additional reservoir was impounded for the Languedoc canal by the Lampy gravity dam. Although its almost rectangular cross section with a base width of 70% of the dam's height was stable in itself, several short buttresses were built against the downstream face.

Bold Gasco dam ends in failure

A similar enterprise near Madrid conceived as part of a trans-Iberian canal from the Atlantic Ocean to the Mediterranean Sea was less fortunate than the Languedoc canal (Garrandes 1973). The grandiose scheme was proposed in 1785 by a bank and designed by Carlos Lemaur (died 1785), a French engineer and brigadier in the Spanish army, together with his four sons. In 1787 construction started on the 87 m (!) high Gasco dam on the Guadarrama river, 30 km northwest of Madrid. It was designed as a buttress dam, with sloping decks closing the intermediate spaces on both faces (Fig. 122). To provide additional weight, these spaces were filled with earth, a provision which entailed the ultimate failure of the bold structure. After a heavy rainfall in May 1799 the saturated earthfill pushed out the downstream slab over four intermediate spaces, whereupon, at a dam height of 56 m the already flagging enterprise came to a standstill (Fig. 123).

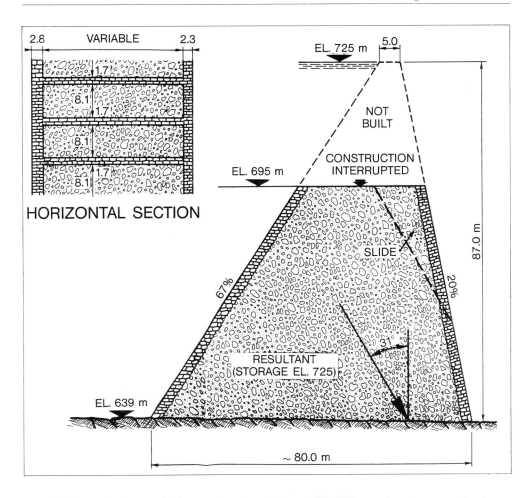

HORIZONTAL SECTION

Figure 122. Vertical and horizontal sections of the Gasco navigation dam northwest of Madrid in Spain (after Garrandes 1973).

Figure 123. Downstream view of the abandoned Gasco dam, showing the pushed out part of the deck near the crest (photo by the author).

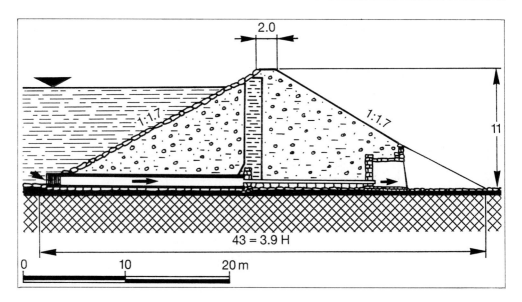

Figure 124. Cross section of the Blackbrook navigation dam near Leicester in Great Britain (after Binnie 1987).

Unlucky Blackbrook dam

Succesfull Rudyard embankment

Construction details

One of the first navigation dams built during the Industrial Revolution in Great Britain, the Blackbrook embankment completed in 1797 for the Leicester waterway was also plagued by misfortune (Binnie 1987). No later than two years after completion it was washed away by a flood and then again in 1801, whereupon the site was abandoned for over one century. The dam had been designed by William Jessop (1745-1814) with a central puddle clay core, as developed for the abovementioned ornamental lake dams. Both faces sloped rather steeply at 1:1.7 (Fig. 124). Jessop subsequently designed the successful Upper Chelburn, Hollingworth and Blackstone dams for the Rochedale canal from Manchester over the Pennine hills to the industrial areas in the East. However, his recommendation to provide the dams with puddle clay cores was not adhered to.

A puddle clay core was prescribed by John Rennie (1761-1821) for the Rudyard embankment completed in the same year as the Blackbrook dam. According to his specifications, which were to become exemplary for more than 100 years, the core was 1.8 m wide at the top of the 10.5 m high dam and 3.7 m wide at its base. Within its base, there was a 3.7 m deep by a 4.6 m wide trench filled with puddle clay. The crest width of the embankment was 12.2 m or 116% of the dam height and it sloped 1:1.7 on the upstream and 1:2.0 on the downstream faces. Placement of the shells was in layers of 0.6 m thickness dipping towards the core, so that both would lean against it, while wet conditions were maintained around the core. A really sturdy structure! This was also true for the spillway at the left end of the dam which consisted of a small arch dam between two mighty wing walls of superbly executed ashlar masonry (Fig. 125).

A third dam, completed in 1797 (Table 28) near Slaithwaite for the Huddersfield canal over the Pennine hills to a height of 21 (later 22 m), was to remain the highest in Great Britain for some 40 years. It was designed by Benjamin Outram (1764–1805), together with the Tunnel End and Diggle dams at each end of the 5 km (!) long Standedge summit tunnel of the same canal. Like the lower structures, the Slaithwaite dam probably had no core and leaked badly upon impoundment. Moreover, its outlet broke under the settling of the embankment and had to be replaced by a new one. Also, the embankment slopes, originally 1:2.5 on both faces, were gradually flattened to about 1:4. Outram was also responsible for the Combs dam southeast of Manchester, completed in 1805, the year of his death at the young age of only 41 (Ferguson et al. 1979).

Troublesome record embankment

5.7 FLOOD PRODUCING AND FLOOD RETENTION DAMS

A special form of transportation on water was the fluming and rafting of wood in creeks and rivers respectively. In creeks natural floods were usually utilized for fluming. But when wood consumption was high and continuous, for example in smelting or salt-works, artificial floods were caused when needed. These were brought about by means of dams with relatively large outlets for the rapid release of the reservoir content. Originally, such fluming dams were built entirely of timber and little remains of them in either descriptions or pictures. From about the end of the 17th century onwards, some fluming dams were built of masonry and the ones on which sufficient data are available, are listed in Table 29

Wood transportation by fluming

Fluming dams

Table 29. European fluming dams built of masonry (of which dimensions are known).

Year of completion	Name of dam	Distance in km from city/country	Height (m)	Length (m)	Reservoir capacity (million m^3)
1695	Joux Verte[1]	30 SE Lausanne/Switzerland	8	30	small
1756	Maria Theresia	30 SW Vienna/Austria	6	?	0.1
1769	Belčna	44 SW Ljubljana/Slovenia	18	35	?
1772	Idrijska	46 SW Ljubljana/Slovenia	14	41	0.2
Ca. 1772	Smrečna	36 SW Ljubljana/Slovenia	12	27	?
Ca. 1775	Putrihova	46 SW Ljubljana/Slovenia	15	44	?
1798	Schwarzenbach[2]	40 S Karlsruhe/Germany	12	25	1.0
1812	Ovčjaška	? W Ljubljana/Slovenia	16	35	?

[1]Curved in plan; failed 1945. [2]With lateral embankments.

Joux Verte arch dam

(Breznik 1984, Hafner 1977-1978, Schnitter 1992). The oldest of these fluming dams near Joux Verte in southwestern Switzerland was particularly remarkable in that it was strongly curved in plan to a central angle at the crest of 120° (Fig. 126) (Schnitter 1992). Its 3.5 m wide rectangular cross section had an earth core of 0.8 m width akin to that of the Roman arch dam at Baume in southeastern France and the water retaining walls of the recent embankments near Versailles (see above) were many Swiss served as mercenaries. Moreover, its upstream face was lined with clay contained by a wooden wall, which was prolonged by a similar floor in front of the bottom outlet.

Outlet operation

The outlet was 3.5 m wide and 1.7 m high and closed by wooden double doors that were bolted with a horizontal beam. The latter was supported by a downstream cross beam. By letting a weight fall on it, the horizontal beam could be removed and the doors opened instantaneously. Downstream, a 5.9 m long wooden channel directed the water jet against the piles of logs about 1 m long, intended to be flumed down to the main valley. After fluming had been discontinued, the interest in the Joux Verte dam was no longer maintained. Consequently, in 1945, i.e. 250 years after its construction, a flood succeeded in tearing away its central part. The two wings were protected in 1982/1983 against further decay.

Austro-Hungarian fluming dams

During the reign of the Austro-Hungarian empress Maria Theresia (1740-1780), a substantial fluming dam named after her was built in the Viennese forest (Hafner 1977-1978). It consisted of two ashlar masonry walls with an earth core in-between and had two outlets of 1.7 by 1.7 m clear opening. This was also the case at the slightly younger Belcna dam in western Slovenia, a region which, at that time was part of the Austro-Hungarian empire (Breznik 1984). Its two outlets were even 5.1 m high and had a width of 3.7 m (Fig. 127). The dam's cross section of solid ma-

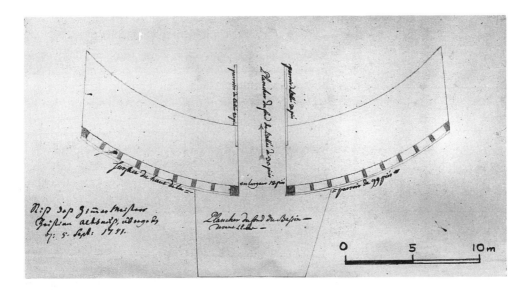

Figure 126. Plan from 1751 of the Joux Verte fluming dam in southwestern Switzerland (Bernese state archive BV 379, p. 1616).

Figure 127. Downstream view of the Belčna fluming dam in western Slovenia (photo A. Vogel, Vienna).

sonry, was trapezoidal, with a vertical upstream face, a 6.8 m wide crest and a base width of 12.4 m or 68% of the height. The engineer in charge was the surveyor Jožef Mrak (1715-1785) from Idrija, 36 km west of Ljubljana, who was also responsible for an additional four fluming dams of similar design in that area (Table 29).

Naturally, at the end of the fluming stretch, the logs had to be recovered. Diversion racks were usually built diagonally across the creeks for that purpose. In some instances, however, the racks were installed on the top of a dam, thus raising the water level to an elevation convenient for the recovery of the logs. Fine examples of such an installation were the two stone filled timber crib dams built in 1689-1690 just above the confluence of two creeks near Müh-

Wood recovery dams

letal in the Gadmen valley, 68 km southeast of Berne/Switzerland, by one Melchior Gehren (Fig. 128) (Schnitter 1992).

Flood retention

In the era of postmedieval Europe until the end of the Napoleonic wars (1815) very few flood or debris retention dams were built in contrast to the many flood producing fluming dams erected during the same period. One of the oldest flood retention dams was the Pontalto dam in then Austrian Southern Tyrol (Noetzli 1932). The construction began in the year 1611 in a narrow gorge of the Fersina creek. The specific purpose was to

Pontalto arch dam

protect the city of Trento against its floods. The first stage, completed in the following year, was 5 m high and slightly curved

Figure 128. Downstream view of the timer crib dams carrying the log recovery racks at the end of two fluming stretches in central Switzerland (engraving after a painting of Caspar Wolf).

in plan with a central angle of 20°. In 1749-1752 it was heightened to 17 m and the central angle at the 10 m long crest thereby attained now 40°. The dam's constant width was doubled from 2 to 4 m. Later, between 1824 and 1887, it was heightened to almost 40 m in order to keep up with the progressive silting-up of the reservoir. The structure thus served partly as a debris retention dam, a technique that was tested in Tyrol since the 16th century (Lanser 1962). It was the South-Tyrolean Josef Duile (1776-1863) who published the first textbook on debris control in mountain creeks in 1826, describing amongst other methods the one of check dams. Many of these were built afterwards.

Debris retention

5.8 TECHNOLOGY TRANSFER TO THE COLONIES

Along with their expansion to overseas territories the European conquerors of the 16th century introduced their hydraulic technology to many countries. In Central and South America, however, the Spaniards encountered civilizations that, under quite similar natural conditions, had developed a long tradition of hydraulic engineering, including the construction of dams (see above). By and large, they could therefore carry on with what had already been done, just introducing new techniques here and there, such as e.g. arched bridges for aqueducts, water wheels for mills, masonry for dams etc. It was mainly the clergy who initially took these matters in hand for the supply of water to the newly founded monasteries and their fields.

Improvement of existing techniques in Latin America

Role of clergy

Thus, the Augustinian brother Diego de Chávez Alvardo (1508-1573), nephew of Pedro Alvarado, the lieutenant of Mexico's conqueror Hernando Cortez, was responsible for the huge Yuriria irrigation reservoir containing 221 million m³, which he impounded in a swampy volcanic crater 230 km northwest of Mexico City in 1548/1550 (Díaz-Marta 1981). A 12 m high gravity dam closed it off completely, whereas a 6 km long diversion canal from the Lerma river was built to fill it. Thereafter, many fine aqueducts were erected all over Mexico (Schnitter 1982), but no more dams were built until about 1730 (Rec. Hydr. 1976).

Huge Yuriria reservoir

In the meantime, an intricate system of reservoirs and canals was constructed from 1573 to 1621 near Potosí, 415 km southeast of La Paz in Bolivia, where the Spaniards had discovered particularly rich silver mines (Fig. 129) (Rudolph 1936). They were located between 4200 and 4800 m a.s.l., in an area where the scarce yearly precipitation of about 600 mm falls within three months. Some 30 reservoirs with a total content of approximately 6 million m³ were therefore required to continuously drive the 132 ore grinding mills

Potosí power dams

Figure 129. Map of the mining
reservoirs and canals (broken
lines) near Potosí in Bolivia
(after Rudolph 1936).

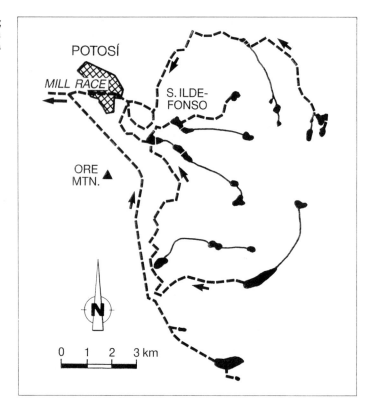

Dam design

San Ildefonso failure

Resumption of dam
building in Mexico

of some 450 kW total capacity. The dams impounding the reservoirs were gravity structures with earth cores of the ancient type that had been revived a few years earlier for the Ontigola irrigation dam in the Spanish royal gardens of Aranjuez (see above). As in the latter's final state, the largest, about 8 m high Potosí dams had three masonry walls and two cores as well as a base width of up to 150% of their height. In 1626 the 8 m high San Ildefonso dam (nearest to the city) failed and released some 0.4 million m^3 of water, drowning about 4000 people, irreparably destroying 60% of the mills and damaging another 35%. Potosí never completely recovered from the disaster, despite the fact that the dam was speedily repaired and most of the important mills rebuilt.

As mentioned above, dam building was resumed in Mexico around 1730 and the data of the more important structures are summarized in Table 30 (Díaz-Marta 1981, Hinds 1932, Rec. Hidr. 1976). The first five were all located near Aguascalientes, 430 km norhtwest of Mexico City. At their toes they powered mills and served for irrigation purposes as well. The Saucillo dam was still a rectangular gravity wall with a base width of just 27% of the height (Rec. Hidr. 1976). It probably owed its stability to its two halves forming in plan an arch like V-shape pointing upstream.

Table 30. Colonial dams in Mexico.

Year of completion	Name of dam	Distance in km from city	Type	Height (m)	Length (m)	Reservoir capacity (million m³)
1550	Yuriria	110 SE León	Gravity	12	?	221.0
1730/1940	Saucillo	39 N Aguascalientes	Gravity	11	175	6.0
Ca. 1750	Arquitos	10 NW Aguascalientes	Buttress	20	219	?
Ca. 1750	San Blas	33 N Aguascalientes	Buttress	24	177	1.0
Ca. 1750	San José	17 NW Aguascalientes	Buttress	11	?	?
1760/1971	Natillas	54 N Aguascalientes	Buttress	12	100	0.5
1765/1936	Huapango	90 NW Mexico City	Gravity	14	840	1.9
1765/1940	San Antonio	150 NW Mexico City	Gravity	11	150	3.0
1772	Aguacate	130 SE León	Gravity	12	260	1.5
1800	Ñadó	110 NW Mexico City	Gravity	26	180	7.0

In the centre there was a massive block which contained the bottom outlet.

The remaining four dams near Aguascalientes were all of the buttress type. This had been introduced by Becerra for power dams in southwestern Spain in 1565 and reached its peak at the Feria dam in 1747 (see above). The three cross sections in the upper part of Figure 130 show that the Aguascalientes dams also consisted of more or less rectangular water retaining walls propped up by irregularly spaced buttresses (Hinds 1932). The youngest buttress dam near Natillas was unusual in that it was built of cobble masonry (Fig. 131). Thereafter the buttress type was abandoned in favour of gravity dams, at least two of which have triangular cross sections (bottom of Fig. 130) (Rec. Hidr. 1976). Unfortunately it remains unknown who their ingenious builder was. He had adopted the modern shape almost one century before it was developed in France (see below).

The situation encountered by the European conquerors in the east of North America was quite different from that of Latin America since – due to the climate and low population densities – there was neither need for irrigation nor for water supply works. Like in Latin America, the water power technology had to be imported. This, among other things was carefully prepared in advance by the 'Mayflower' expedition to New England in 1620 (Fleishman 1978). Only 12 years after the first landing the Charles river was dammed in 1632 about 10 km northwest of Boston in Massachusetts and a grist mill was built on the site two years later (Fig. 132). During the ensuing $2^1/2$ centuries, the 127 km of river course from its mouth to Lake Echo and a total head of 107 m was developed by two dozen dams. These were made of either wood or stone, and more often, a combination of both. Consequently the need for maintenance was constant. On the whole, only a few

Buttress dams

First triangular gravity dams

Development of water power in North America

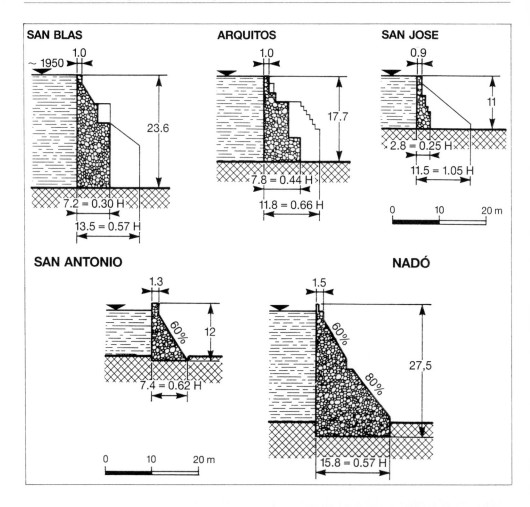

SAN BLAS

~ 1950

1.0

23.6

7.2 = 0.30 H

13.5 = 0.57 H

ARQUITOS

1.0

17.7

7.8 = 0.44 H

11.8 = 0.66 H

SAN JOSE

0.9

11

2.8 = 0.25 H

11.5 = 1.05 H

0   10   20 m

SAN ANTONIO

1.3

60%

12

7.4 = 0.62 H

0   10   20 m

NADÓ

1.5

60%

80%

27,5

15.8 = 0.57 H

Figure 130. Cross sections of five 18th century dams in Mexico (after Hinds 1932 and Rec. Hydr. 1976).

Figure 131. Downstream view of the Natillas buttress dam north of Aguascalientes in Mexico (photo by the author).

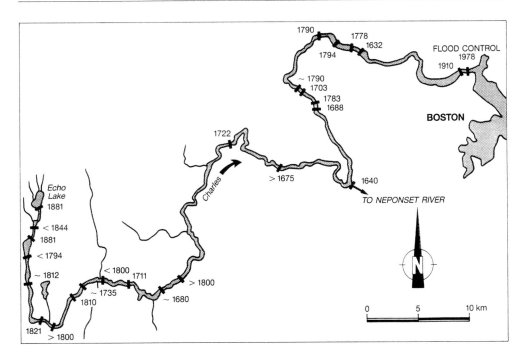

major washouts of dams were recorded, a tribute to the artisans who built them.

The power potential of many rivers and creeks in North America was developed in similar fashion. The earliest printed review of the technology used was 'The Young Mill-Wright and Miller's Guide', first published in 1795 by the wheelwright Oliver Evans (1755-1819) (Evans 1795). In his publication he described the two principal types of timber dams, i.e. the stone filled crib (Fig. 128) and the log dam. Like the multiple arch dams proposed and built in Spain in the 16th and 18th centuries (see above), the latter sloped upstream thus using the stabilizing effect of the water load on the upstream face (Fig. 133). Moreover, it minimized the danger from the scouring action of overflowing water by having its crest as far downstream as possible.

In his book, Evans also dealt with masonry dams and recommended that they be curved in plan. This recommendation was adhered to by lieutenant-colonel John By (1779-1836) of the British 'Royal Engineers' after he had been recalled from his retirement and sent to Canada to build the Rideau barge canal from Ottawa to Kingston on Lake Ontario between 1828 and 1831 (Legget 1955, 1972). Also, several weirs and dams had to be built along the canal. Probably as a consequence of the failure of his straight gravity dam at the Hog's Back in Ottawa in 1829 By gave the Jones Falls dam a strong curvature in plan with a central angle at the crest of 115° (Fig. 134). In any case, that failure was the more

Figure 132. Map of the Charles river west of Boston in Massachusetts with location of the mill dams built since 1632 (after Fleishman 1978).

Crib and log dams

Rideau barge canal in Canada

Figure 133. North American log dam with sloping upstream face (from Leffel 1881).

Figure 134. Downstream view of the Jones Falls arch dam in eastern Ontario/Canada (photo by the author).

Jones Falls arch dam

likely reason than the knowledge of Evans' book. The 19 m high Jones Falls dam had an almost constant width of 44% of its height and, for arch dams, an unprecedentedly high length-to-height ration of 5.6. The ashlar masonry blocks were set in vertical courses (rather than the usual horizontal ones), so as to more closely follow the dam's curvature. Moreover, clay fill was placed against the upstream face as an additional water barrier.

European dam technology in India

By's mission in Canada was probably one of the last of its nature in North America. In other parts of the British empire however, the 'Royal Engineers' were to play an important role in hydraulic and dam constructions for much longer. Notably in India, where the introduction of European dam building techniques in the 16th century by the Portuguese and well before the British takeover has been mentioned already. It may have been due to their influence that an earth-backed masonry wall was adopted for the 23 m high

Figure 135. Downstream view of the Mir Alam multiple-arch dam near Hyderabad in central India (photo by the author).

Hussain dam in Hyderabad, 1240 km south of Delhi, in 1575 (Murti 1979). This design had never before been used in eastern Asia. In the West it had been utilized already by the Romans and was revived just 15 years earlier for the Granjilla 2 irrigation dam in Spain (see above). In 1730 a similar, 30 m high dam was built in Jaisamand, 590 km southwest of Delhi (INCOLD 1979).

The extraordinary Mir Alam (or Meer Allum) multiple-arch dam completed in 1804 for the water supply of Hyderabad unquestionably was the work of a member of the British 'Royal Engineers', one Henry Russle (Fig. 135) (Eng. Rec. 1903). The structure, up to 12 m high, consisted of 21 semicircular, vertical arches of constant thickness but with variable spans of up to 51 m. As with the exception of the abovementioned, much lower and sturdier structures of the 18th century in Spain, no precedents are known and since almost 100 years went by until another similar work was completed in Australia (and 120 years until the arch span of 51 m was again attained at a multiple-arch dam in Arizona, see below), the Mir Alam dam must be considered as one of the rare, true strokes of genius!

Earth-backed masonry dams

Extraordinary Mir Alam multiple-arch dam

5.9 SUMMARY

The description of dam constructions in medieval and postmedieval Europe up to the end of the Napoleonic wars (1815) has proceeded within the framework and according to their specific purposes. It was not so much a desire to illustrate the latter's great variety, but rather because, originally, the builders of dams destined to serve a certain use, knew and cared little about the structures

Improved
communication among
dam designers

built for other purposes. Moreover, communication was difficult in those earlier times! This situation changed gradually with the publication of the first books dealing with dam engineering partly mentioned before (Table 31). An important aspect also was the establishment of engineering schools, like the one in 1747 for the French 'Corps of Bridge and Highway Engineers', and the emergence of societies of civil engineers, as for example the one in Great Britain in 1771, which in 1831 and 1836 respectively began to regularly publish technical papers. But even before many reports on individual projects had been printed, particularly in Great Brittain (Skempton 1977).

Review of five centuries
of European dam
building

For a review of what had actually been built ('The proof of the pudding lies in the eating'), in the way of dams of over 20 m of height in Europe or by Europeans for the variety of purposes dealt with above, they are listed chronologically in Table 32 (excluding the unsuccessfull Gasco and Puentes dams in Spain). While some world records in height were broken earlier, the 18th century was marked by a rapid increase of the number of dams exceeding the adopted height limit. This was a foreboding of the acceleration of European technology during the Industrial Revolution and afterwards. With regard to the different structural types, the following may be noted:

Embankments

– Embankment dams: The maximum height was increased by only a few metres over the limit already reached in 460 AD at the Paskanda Ulpotha dam in Sri Lanka (see above). The most important innovation was the introduction of the central impervious core in the middle of the 16th century.

Gravity dams

– Gravity dams: Some strides were taken in the maximum height achieved (46 m at Tibi against 40 m of the Roman dam at Subiaco/Italy in the first century). But the Roman concept of a rectangular wall was mostly maintained, with only a few, hesitant attempts to use trapezoidal, let alone the correct triangular, cross sections. The use of concrete by the Romans for the dams' interior or as a building material in general, fell into oblivion, while the

Table 31. Early books on dam engineering.

Year first published	Author	Subject	Languages
1547	Dubravius (1480-1553)	Fish ponds	Latin/German/English
1600	J. Taverner (?-1606)	Fish ponds	English
1724/1735	J. Leupold (1674-1727)	Dikes/weirs	German
1736	P.B. Villareal (1670-1740)	Power dams	Spanish
1754/1757	A. Brahms (1692-1758)	Dikes	German
1763	H. Calvör (1686-1766)	Mining dams	German
1779	L. Voch (1728-1783)	Timber dams	German
1786	J.E. Silberschlag (1716-1791)	Dam failures	German
1795	O. Evans (1755-1819)	Mill dams	English

Table 32. Successfull European dams of over 20 m in height (up to 1815).

Year of completion	Name of dam	Country	Height (m)	Type	Purpose
13th cent.	Almonacid	Spain	25	Gravity	Irrigation
1384–1586		Spain	23	Gravity[1]	Irrigation
1594	Tibi	Spain	46^2	Gravity[1]	Irrigation
1614	Dolná Hodruša	Slovakia	22	Embankment	Power
1640	Elche	Spain	23	Arch	Irrigation
1675	St. Ferréol	France	36^2	Embankment	Navigation
1704	Arguis	Spain	23	Gravity	Irrigation
1722	Oder	Germany	22	Gravity	Power
1740	Large Richnawa	Slovakia	23	Embankment	Power
1744	Rozgrund	Slovakia	30	Embankment	Power
1745	Horná Hodruša	Slovakia	22	Embankment	Power
Ca. 1750	Arquitos	Mexico	20	Buttress	Pow./Irrig.
Ca. 1750	San Blas	Mexico	24^2	Buttress	Pow./Irrig.
Before 1770	Klinger	Slovakia	22	Embankment	Power
1776	Relleu	Spain	29	Gravity	Irrigation
1779	Počúvadlo	Slovakia	23	Embankment	Power
1788	Large Taul	Romania	28	Embankment	Power
1797	Slaithwaite	Great Britain	21	Embankment	Navigation
1800	Ñadó	Mexico	26	Gravity	Irrigation
1806	Valdeinfierno	Spain	30	Gravity	Irrigation
1811	Couzon	France	35	Embankment	Navigation

[1]Curved in plan. [2]World record for type.

Table 33. European reservoir systems.

Period of construction	Name of system	Country	Purpose	Reservoirs Number	Capacity (million m^3)	Figure
1404–1844	Freiberg	Germany	Min. Power	20	5.7	–
1530–1600	Třeboň	Czechia	Pisciculture	~120	~60.0	107
1535–1808	Harz	Germany	Min. Power	110[1]	~10.0	96
1560–1839	Istanbul	Turkey	Water supply	7	2.3	116
1573–1621	Potosí	Bolivia	Min. Power	~30	6.0	129
1614–1800	Banská Stiav.	Slovakia	Min. Power	54	~7.0	99
1678–1685	Versailles	France	Water supply	~30	8.0	119
1797–1830	Huddersfield	Great Britain	Navigation	8	1.4	–
1799–1857	Rochdale	Great Britain	Navigation	7	7.0	–

[1]Not in operation simultaneously.

construction techniques essentially remained the same: pick and shovel.

– Buttress dams: The Roman concept of propped up simple walls was developed to the astonishing height of 24 m at the San Blas dam in Mexico. Also, the multiple-arch dam idea was revived in some low dams in Spain and, in 1804, at the extraordinary

Buttress dams

Indian dam Mir Alam, designed by an ingenious member of the British 'Royal Engineers'.

Arch dams

– Arch dams: The few true structures of the type built in Europe up to the year 1815 did not match the achievements attained in Iran under the Mongolians during the 14th century.

Reservoir systems

The preference for embankments in northern Europe on one hand and that for masonry structures in the Romanized south, noticed in connection with irrigation dams on the other, applied to projects for all purposes. Similarly, the reservoir systems, which were now built for the second time after those in ancient Sri Lanka (see above), were mostly located in northern Europe (Table 33). Although the amounts of stored water were generally much smaller in the European systems, the number of dams involved and the speed of their construction (up to four per year at Versailles) were sometimes impressive.

CHAPTER 6

Evolution of modern dam technology

As just mentioned the Industrial Revolution accelerated considerably in the wake of the French Revolution and the Napoleonic wars and engulfed all of western Europe and North America. Since it was partly based on the practical application of science, technical universities and engineering schools were established in many countries. The hitherto individualistic craftsmanship of engineering thus became a profession based on common scientific principles. Due to the rapid growth of easily available knowledge, engineers inevitably were compelled to specialize in certain fields, such as for example in hydraulic engineering or just dams. Independently of any given project's purpose, dam technology subsequently even needed a subdivision, usually in accordance with structural types or elements. This subdivision is adopted in the present chapter, beginning with the embankment dams as they represented the most frequent type and, moreover, because they were the first ones to be investigated scientifically – alas without immediate consequence!

Impact of Industrial Revolution

Specialization of dam engineering

6.1 EMBANKMENT DAMS

6.1.1 *First attempts at rationalization*

In 1717, a senior engineer of the French 'Corps of Bridge and Highway Engineers', Henri Gautier (1660-1737), published a treatise on bridges including earth retaining walls, of the kind that had been used in the dams for the Versailles park (see above) (Skempton 1979, 1985). He experimentally determined the natural slope of repose of soils and recommended that the mean thickness of a retaining wall with an inclination of the air face of 20%, should be 25% of its height. Thirty three percent were postulated in 1729 by brigadier and professor of mathematics Bernard Forest de Bélidor (1671-1761), who later published four volumes on

Studies on slope stability

Figure 136. Charles A. Coulomb (1736-1806).

Figure 137. Upstream view and cross section of the Cercey dam 42 km west of Dijon in France, which experienced several slips in its down- and upstream faces; in broken lines one of the counterforts used to repair the upstream face (photo by the author, section after Ziegler 1900-1925).

hydraulic engineering (excepting dams). It was to become a standard reference for civil engineers not only in France but also in Germany and Great Britain. The first director of the school for the French 'Corps of Bridge and Highway Engineers', Jean-Rodolphe Perronet of Swiss origin (1708-1794), initiated the engineering study of slope stability in excavations as well as embankments with a memoir published in 1769. He already advocated test borings (either driven or by auger) and pits preceeding design and construction of earthworks.

Four years later, the French military engineer Charles A. Coulomb (1736-1806) (Fig. 136) read his historic paper on the method of limit equilibrium analysis in soil mechanics to the (then still) 'Royal Academy of Sciences'. It appeared in print in 1776 and established the basic concept whereby the sliding resistance of soils depends on their cohesion, which is proportional to the area involved, and furthermore, their internal friction, which is a function of the acting normal forces. His formulae for the critical height of just stable cuts in clayey soils were extended to slopes in 1820 by Jacques-Frédéric Français (1775-1833), professor of the military college of Metz in northeastern France.

It was, however, Alexandre Collin (1808-1890) who substantiated Coulomb's theories by field observations and shear tests

(Skempton 1946, 1949). After graduating from the Polytechnic School in Paris he entered the 'Corps of Bridge and Highway Engineers' and was appointed to the works then resumed on the Burgundy navigation canal near Dijon in eastern France. Here, he started to investigate slips, which occurred in various canal works, as e.g. several times up to 1842 on both faces of the Cercey storage embankment completed in 1836 (Fig. 137).

By 1840 Collin had determined the circular shape of slides in clayey soils as well as the great significance of the water content of such soils for their cohesive strength. Although he thus came close to modern concepts, his findings fell into oblivion for some 70 years. On the one hand, more authoritative researchers advocated at about the same time that cohesion could be discounted. On the other hand, the emerging science of statics led French engineers to give preference to masonry dams as it was already possible to analyse these in greater depth (see below). As a result, the construction of embankment dams was virtually discontinued in France.

Collin close to modern concepts

6.1.2 *Heydays of empiricism*

Progress in Great Britain was meanwhile following the opposite direction of the one in France. As already noted, there was a great upsurge in reservoir construction in order to satisfy the water demand of urban populations, which, in the wake of the Industrial Revolution, grew rapidly. All were impounded by embankments, the design and construction of which was based entirely on the empirical knowledge gained with the ornamental and navigation dams (see above). The first important water supply structure was the Glencorse dam built between 1819 and 1824 some 15 km south of Edinburgh, to compensate the local mill proprietors and landowners for the water diverted to the city (Binnie 1987). It was designed by the famous British engineer Thomas Telford (1757-1834) with the assistance of Rennie, who had built the successful Rudyard navigation dam mentioned above. Like the latter, the Glencorse embankment had a puddle clay core but somewhat flatter outer slopes (Fig. 138). Its height was increased in 1848 to

Boom of water supply dams in Great Britain

Figure 138. Cross section of the Glencorse water supply dam south of Edinburgh (after Binnie 1987).

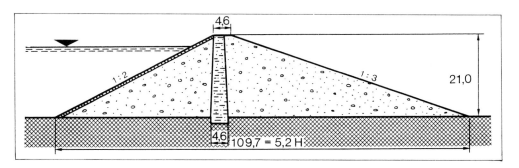

23 m, a mark which had been surpassed already in 1837 at the Entwistle dam.

Record height earth dams

Located some 20 km northwest of Manchester the Entwistle reservoir was originally built by some millers for storing surplus runoff during the rainy winter months and used only later for water supply (Binnie 1987). The design for the 33 m high structure was prepared by the local surveyor Thomas Ashworth and checked by Jesse Hartley (1780-1860), chief engineer of the port of nearby Liverpool. In 1840, the dam was heightened to 38 m and thus became the highest embankment in the world until 1882, when the Upper Barden dam was completed some 60 km north of Manchester. This 42 m high water supply structure was designed by Alexander R. Binnie (1839-1917) and remained the highest embankment in Great Britain until World War II (1939-1945).

Proliferation...

Until 1930, i.e. the advent of modern soil mechanics (see below), some 260 embankments of over 15 m in height were built in Great Britain in slightly more than one century, mostly for water supply (top of Fig. 139). As mentioned before, the technology was also exported to the colonies, particularly to India (incl. Bangladesh and Pakistan), where some 80 such structures were completed during the same period to either serve irrigation and/or water supply. Many of these dams were designed and supervised by just a handful of engineeers, the most prolific of which were James Leslie (1801-1889), the cousins John T. and John W. Leather (1804-1885 and 1810-1887), Thomas Hawksley (1807-1893) as well as John F. (La Trobe) Bateman (1810-1889) with his partner George H. Hill

...and standardization of embankments in Great Britain

(1827-1919). This resulted in a strong concentration of knowhow and a remarkable degree of standardization. It also may explain why the British embankment dams built up to 1930, although based completely on experience, only suffered four failures, the most disastrous of which were the collapses of the Bilberry embankment, 36 km northeast of Manchester, in 1852 (81 deaths) and that of the Dale dike 40 km east of Manchester in 1864 (238 deaths) (Binnie 1981).

Many embankment failures in the USA

In contrast, the safety record of the embankment dams of more than 15 m height, built in the USA from around the middle of the 19th century was appalling (bottom of Fig. 139) (USCOLD 1975, 1988). Of some 380 embankments completed by 1930, over 9% failed, three fifths of which within the first ten years of their operation, mostly due to defects in the embankments. The main cause for the later failures was overtopping by floods. While this might be considered typical for newly settled territories with only few, if any, hydrological records, the structural failures must be attributed to empiricism not as yet solidly based on tradition as well as to a considerable willingness to try new ideas. Notwithstanding the losses, these full-scale 'experiments' were to benefit worldwide

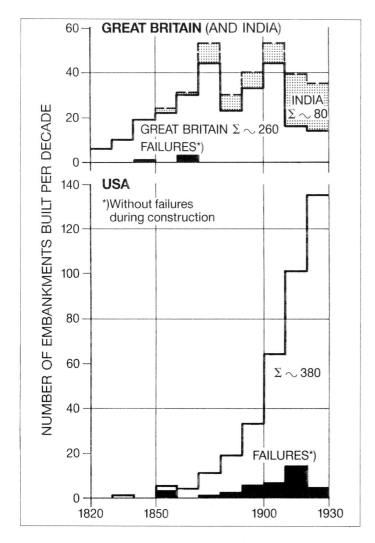

Figure 139. Number of embankment dams of over 15 m height built up to 1930 in Great Britain and India (incl. Bangladesh and Pakistan) as well as in the USA.

dam engineering, not least because of the candid reporting in the American technical press.

6.1.3 *Trying new ideas*

One of the earliest American embankments with over 15 m of height, built in three stages between 1838 and 1852 for Pennsylvania's 'Main Line' navigation canal near South Fork, 100 km east of Pittsburgh, already displayed a novel internal zoning of the materials (centre of left part of Fig. 140) (McCullough 1968). The young state engineer William E. Morris designed almost the complete upstream half of the cross section of puddled earth, while, separated from it by a transition zone of slate, the downstream shell

Novel American
embankment design

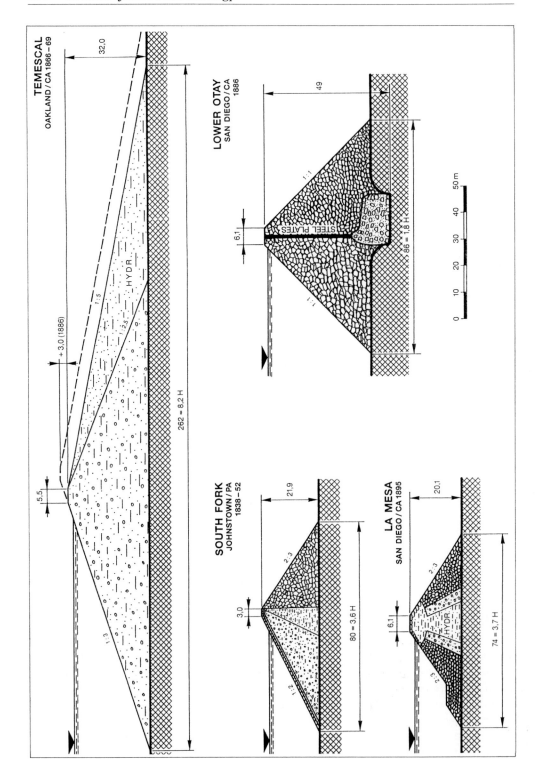

consisted of loose rocks. Such rockfills were to become very important for embankment dam construction in the USA, and later, on a worldwide scale. After a blowout along its outlet in 1862, the South Fork dam remained unused until 1880, when it was bought and repaired by a fishing and hunting club. At the end of May 1889 it was overtopped by a flood, a disaster exacerbated by the obstruction of the spillway with fish-screens. After about three hours, the dam burst and the ensuing flood killed 2209 people.

In the meantime, a few embankments had been built based on the British standard design with a thin puddle core such as, for example, in the years from 1864 to 1871, the 36 m high Druid Hill dam in Baltimore/Maryland (Wegmann 1888-1927). From 1866 onwards, three similar though smaller dams were constructed according to the designs of Hermann F.A. Schussler, a native of Switzerland (1842-1919). They served the supply of water to San Francisco/California, were heightened several times and successfully withstood the catastrophic earthquake of 1906. The same was true for the 32 m high Temescal dam built from 1866 to 1869 for the supply of water to nearby Oakland (Fig. 141). It was designed by the Canadian-born engineer Anthony Chabot as a homogeneous embankment, with relatively flat outer slopes. Almost as an afterthought, Chabot had the downstream face flattened still further during the last year of the construction period. Drawing on his experience in mining, he used the hydraulic method for the additional fill. Earth and gravel were excavated by sluicing by means of strong water jets and transported to the dam site in pipes or flumes. The material was then placed by settling from the outside towards the centre of the dam (upper part of Fig. 140).

Another innovation emanating from the Californian mining

See Page 160

Figure 140. Cross sections of four early embankment dams in the USA (after Wegmann 1888-1927).

British standard design in the USA

Introduction of hydraulic fill

Figure 141. Modern upstream view of the much transformed Temescal dam 18 km east of San Francisco in California (photo by the author).

Pure rockfill dams

industry – that had become very active after the goldrush of 1848-1849 – was the pure rockfill dam, developed in the Sierra Nevada mountains for storing the water needed for the hydraulic sluicing process. The first rockfill dam for irrigation purposes was built on the Otay creek southeast of San Diego in California (right in lower part of Fig. 140) (Wegmann 1888-1927). It consisted of a simple heap of loose rocks sloping on both faces with their natural angle of repose impermeabilized by a central steel core. With its height of 49 m, it was the highest embankment dam in the world at the time of its completion 1886 (Table 34) (Creager et al. 1945, ICOLD 1973, 1984, Wegmann 1888-1927). Twenty years later it was washed away by a flood (USCOLD 1975, 1988). Other, similar rockfill dams in the southwestern USA were made watertight by upstream decks of timber, steel or concrete. This allowed further reductions of their volume since its mass now fully served to withstand the water load.

Rockfill dams with impervious core

An interesting combination of both techniques derived from the mining industry, i.e. the hydraulic fill process and rockfill, was achieved in 1895 by Julius M. Howells at the 20 m high La Mesa dam (now flooded by Murray dam) in northeastern San Diego (bottom of left part of Fig. 140) (Wegmann 1988-1927). The use of earth cores in rockfill dams provided on both sides with appropriate transition zones was to become very popular. Particularly so after the hydraulic fill process gained a bad reputation and was no longer used due to a number of spectacular construction accidents in the course of which the liquid cores burst. Another disadvantage of hydraulic fills was the fact that they retained their potential for

Table 34. Record height embankments in the USA (up to 1930).

Year of completion	Name of dam	Distance in km from city/state	Type[1]	Height (m)	Length (m)	Reservoir capacity (million m^3)	Purpose
1836	Up. Tumbling	110 W Philadelphia/Pa.	E	18	174	0.9	Water s.
1852	South Fork[2]	100 E Pittsburgh/Pa.	ER	22	256	18.5	Navig.
1859	French	130 NE Sacramento/Calif.	R	30	61	17.1	Irrig.
1869/1886	Temescal	18 E San Francisco/Calif.	HoE	32/35	198	0.6	Water s.
1871	Druid Hill	Baltimore/Md.	ECE	36	914	1.5	Water s.
1875/1892	Chabot	27 E San Francisco/Calif.	HoE	40/43	152	21.8	Water s.
1886	Lower Otay[3]	30 SE San Diego/Calif.	SCR	49	172	?	Irrig.
1914	Swift[4]	190 N Helena/Mont.	CoDER	57	226	37.0	Irrig.
1922	Arrowhead	100 E Los Angeles/Calif.	E	58	219	5.9	Water s.
1924	Stone Canyon	300 NE San Francisco/Calif.	E	67[5]	351	12.8	Water s.
1925	Dix	260 NE Nashville/Ky.	CoDR	87	311	222.0	Power

[1]C = Core, Co = Concrete, D = Deck, E = Earth, Ho = Homogeneous, R = Rockfill and S = Steel. [2]Failed 1889 by overtopping. [3]Failed 1916 by overtopping. [4]Failed 1964 by overtopping. [5]Heightened 5 m in 1955.

liquefaction for a long time and were thus highly susceptible to earthquakes.

6.1.4 *Theoretical breakthrough*

Proliferation of empirical designs

The unfortunate experience with hydraulic fills was typical of the confusion which characterized embankment dam engineering at the end of the 19th and the beginning of the 20th centuries, when empirical rules and various designs proliferated almost as fast as embankment dams themselves. It was high time for their design and construction to become rationalized. An attempt had been made by French engineers one century earlier and scientific principles meanwhile prevailed for all other dam types (see below).

Soil permeability

The problem of soil permeability had already been solved in 1856 by the French water supply engineer Henri P.G. Darcy (1803-1858). He experimentally determined that the velocity of water flowing through sand is directly proportionate to the hydraulic gradient, that is, to the pressure head per unit length of the percolation path (Skempton 1979, 1985). In 1886 the Austrian professor of hydraulics Philipp Forchheimer (1852-1933) found that the flow of water in soils is governed by the general operator formulated by Pierre S. de Laplace (1749-1826). For its solution he developed the graphical trial and error method of the flownet in 1911 (Fig. 142) (Forchheimer 1914). The method had already been discovered four years earlier by the British physicist Lewis F. Richardson (1881-1953) starting from the analogy between the flows of electricity and groundwater, which was later also used directly for model tests (Skempton 1979, 1985).

Flownet

Seepage line

The first determination of the seepage line by measurements in standpipes driven or drilled into an embankment seems to have been undertaken in 1907 by British engineers at the 32 m high Waghad dam, 170 km northeast of Bombay in India (Penman

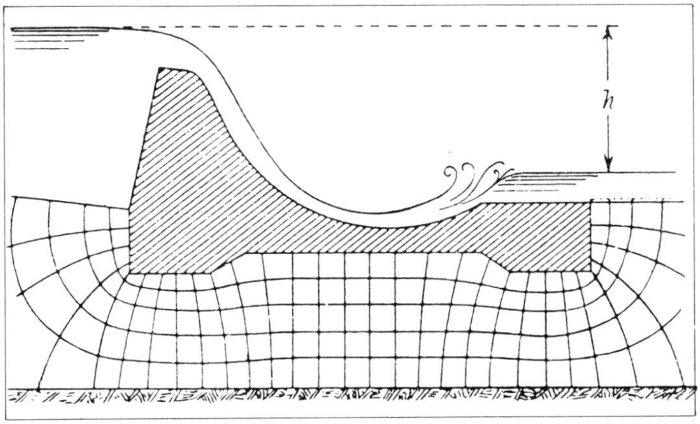

Figure 142. Flownet of the water percolation under a weir on permeable soil (from Forchheimer 1914).

Figure 143. Sliding circle stability analysis of a homogeneous embankment taking into consideration the seepage line E and cohesion (from Felllenius 1926-1939).

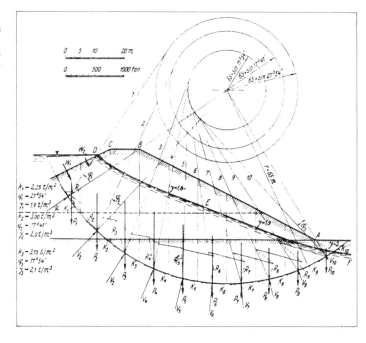

Seepage velocity

Figure 144. Karl von Terzaghi (1883-1963).

1982). Also, colonial engineers first studied the question of the allowable velocity of seepage flow to avoid internal erosion of the soil (piping) (Skempton 1979, 1985). In 1895-1896 John Clibborn (1847-1938) conducted corresponding experiments at the Roorkee laboratory, 150 km northeast of Delhi in India. In 1899 John S. Beresford (1845-1927) demonstrated the effectiveness of filters as a safeguard against piping. Finally, in 1910, William G. Bligh (1846-1923) published empirical values for the permissible gradients based on the experience with actual structures founded on various soils.

Following the failure of a quaywall in Göteborg in southwestern Sweden in 1916, the port authority engineers Knut E. Petterson (1881-1966) and Sven Hultin (1889-1952) rediscovered the sliding circle stability analysis, although in so doing, they considered only the frictional resistance of the soil (Petterson 1955). The inclusion of cohesion was achieved in 1926 by the Swedish professor Wolmar Fellenius (1876-1957) (Fig. 143) (Fellenius 1926-1939).

The single most important milestone in these theoretical developments was, however, the publication in 1925-1926 of the book 'Principles of Soil Mechanics' by Karl von Terzaghi (1883-1963), then professor at the American Roberts College in Istanbul/Turkey and, later, at the Harvard School of Engineering in Massachusetts/USA as well as much sought-after consultant in dam engineering (Fig. 144). As a mechanical, rather than a civil engineer with a particular interest in geology, Terzaghi found the explanation for the consolidation of clays through the dissipation of the water pressure in the soil pores, which is governed by the same law as the diffusion of heat. This also led him to the concept of effective stress, that is, acting stress minus pore-water pressure, which became the cornerstone of modern soil mechanics theory.

In the mid 1930's, engineers of the 'US Bureau of Reclamation' found out that water in standpipes was raising to levels above the bank profile at one of their many earth dams, although the recently completed structure had not as yet been impounded (Penman 1982). They thus discovered the phenomenon of pore-water pressure formation even in unsaturated soils. In the 1940's they tackled the problem theoretically and by laboratory tests as well as by means of many more field measurements. They determined that pore pressure formation depends mainly on the soil's water content, like so many of its other engineering properties, such as its compactness for example.

Surprisingly, the basic relationship between water content and compaction was not satisfactorily established until 1933, when Ralph R. Proctor (1894-1962) of the 'California Department of Water and Power' showed that each soil has an optimum water content at which its dry density under a certain amount of compactive effort attains a maximum. Whether soils should be placed in embankments at optimum water content, or dry or wet of it, became one of the most vigorous arguments in embankment engineering. On the wet side of the optimum water content there was the danger of high pore-water pressures, but when placed too dry a brittle soil could crack, especially if subjected to differential settlements or arching and, possibly, hydraulic fracturing upon impounding. The dilemma had to be resolved for each structure and soil type according to the particular circumstances. The analysis of the intricate stress conditions and hydraulic phenomena within an embankment and its foundation, especially during earthquakes, was greatly facilitated and improved by the introduction of electronics in computation as well as testing and monitoring apparatus.

6.1.5 *Supremacy of embankments*

The theoretical achievements which have been described, together with many more subsequent ones, formed the basis for the vast proliferation of embankment dams around the world. Although they had always dominated the field of the, say, up to 30 m high structures, they penetrated vigorously that of the large dams, especially after World War II (1939-1945) (Fig. 145). The second important basis for this development was that embankments were

Figure 145. Worldwide development of high dam construction since World War II, showing the successive replacement of the gravity and buttress types (PG+CB) by the arch (VA), and, of the latter, by the earth and rockfill dams (TE+ER).

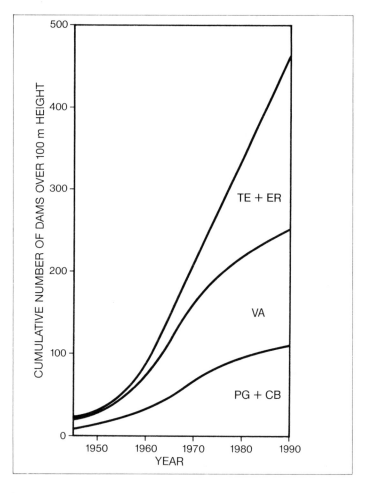

better amenable to the mechanization of their construction. In the 1930's and 1940's, accelerated changes in construction equipment occurred in the USA with regard to capacity and mobility (US-COLD 1988).

Bulldozers attached to caterpillar tractors, known since the beginning of the century, were employed as excavators in addition to their use for spreading the material on the embankment. Power scrapers increased in size to 9 m³ and up to 45 m³ at present (top of Fig. 146). Side-loaders consisting of a plowing blade and a short section of conveyor belt were developed during this period and could load at rates up to 300 m³ per hour. At the same time, power shovels, in use since 1874, were developed with buckets up to about 3 m³ (today about 12 m³, with still larger machines in use in mining).

Progressive increases in the size of rubber tires and the power of traction units permitted the scrapers, wagons and trucks to travel

Mechanization of
construction

on natural soil in the borrow area and on the embankment as well as at relatively high speed on haul roads. Wagons unloaded through their bottom, sidewards or at the rear originally carried some 23 m³ (today up to 55 m³). Trucks usually had a configuration as rear dumps and, by 1940 reached capacities of up to 19 m³ (today about 35 m³). For the compaction of soils, sheepsfoot rollers had been developed in California about 1905 and their use in embankment construction became standard by the 1930's, when their weight reached some 6 tons per metre of roller length (bottom of Fig. 146). After World War II (1939-1945) rubber-tyred rollers and, from about 1955, vibratory rollers were introduced, the latter especially for the compaction of gravel and rockfill, which they greatly improved both qualitatively and quantitatively.

A further reason for the proliferation of embankment dams was the fact that, in contrast to masonry structures, they could also be built on soft soil foundations and consist of a great variety of locally available materials. A new scheme was introduced in the 1930's for their internal zoning by sloping the core upstream, although this may well be regarded as a variation of Morris' 1838 design for the ill-fated South Fork dam in Pennsylvania, which was later adopted for other early embankments in the USA (see above). By shifting the water-barrier upstream a larger part of the downstream shell was obviously available to resist the water load, so that the downstream slope could be steepened (top of Fig. 147). On the other

Figure 146. Horse drawn and modern power scrapers (top) as well as 1930's sheeps foot and modern vibratory rollers (bottom) (photos Morrison-Knudsen Co., Boise/ID, P. Baumann, Los Angeles and T.M. Leps, Menlo Park/CA).

Vibratory rollers

Sloping cores

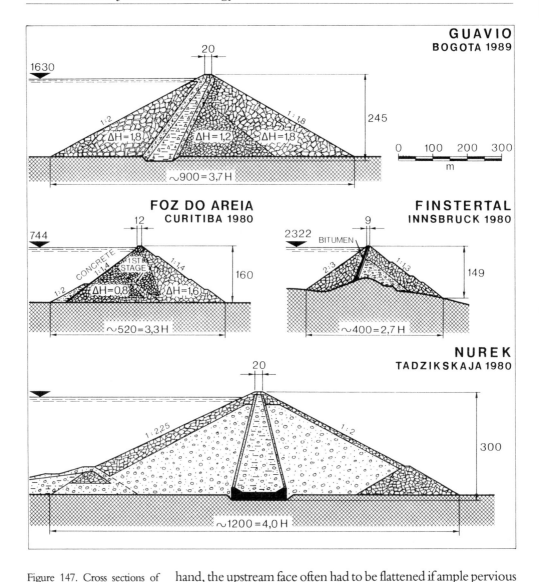

Figure 147. Cross sections of four different types of large modern embankment dams; ΔH = layer thickness.

hand, the upstream face often had to be flattened if ample pervious layers were to provide protection for the impermeable material against the effects of a rapid reservoir draw-down, an aspect Morris had not as yet taken into consideration.

All of the embankment could be used to resist the water load whenever the impermeabilization was arranged on its upstream face, as was the case in the concrete faced rockfill dams already mentioned earlier. This type experienced a great comeback after the introduction of compaction by vibratory rollers reduced their settlement significantly (USCOLD 1988). A further improvement was the adoption since 1967 of nearly monolithic face slabs with only vertical cold joints besides the perimeter joint with multiple waterstops (left centre of Fig 147). In the 1930's the upstream facing

Concrete faced rockfill dams

on a few Algerian and German dams consisted of bitumen, which could adapt to the dams' settlement. Later, this technique was used successfully for the impermeabilization of many rather shallow, off-stream compensating or pumpstorage reservoirs. More recently, bitumen was also adopted for the cores of some large embankment dams (right centre of Fig 147).

Use of bitumen

The rapid increase of height and volume of embankment dams over barely two generations of engineers is summarized in Table 35 (ICOLD 1973, 1984, USCOLD 1988). However, the proliferation of high embankment dams mentioned at the beginning of this section was even more dramatic. It is illustrated by Table 36, which

Ever larger dams

Table 35. World records of embankment height or volume (1930-1989).

Year of comple-tion	Name of dam	Country (state)	Distance in km from city	Type[1]	Height (m)	Embank-ment volume (million m³)	Purpose
(1912	Gatun	Panama	50 NW Panama	H	35	18.0	Navig.)
1931	Salt Springs	USA (Calif.)	210 E San Francisco	CoDR	100	2.3	Power
1932	Afsluit	Netherlands	75 N Amsterdam	E+S	19	63.4	Drainage
1937	Fort Peck	USA (Mont.)	450 NE Helena	H	76	96.1	Multipur.
1939	San Gabriel 1	USA (Calif.)	45 NE Los Angeles	ECR	123	8.1	Flood c.
1948	Mud Mountain	USA (Wash.)	60 S Seattle	ECR	130	1.8	Flood c.
1950	Anderson Ranch	USA (Idaho)	65 SE Boise	ECE	139	7.4	Multipur.
1958	Swift	USA (Wash.)	180 S Seattle	ECGS	156	11.8	Power
1962	Trinity	USA (Calif.)	340 N San Francisco	ECE	164	22.5	Multipur.
1967	Bennett	Canada	750 N Vancouver	ECGS	183	43.7	Power
1968	Oroville	USA (Calif.)	210 NE San Francisco	ECGS	230	59.6	Multipur.
1972	Mica	Canada	440 NE Vancouver	ECGS	242	32.1	Power
1976	Tarbela (Main)	Pakistan	320 NW Lahore	ECGS	143	105.6	Irrig./Pow.
1980	Moreno	Mexico	680 SE Mexico	ECR	261	15.4	Power
1980	Nurek	Tajikistan	45 SE Dushanbe	ECGS	300	58.0	Pow./Irrig.

[1]C = Core, Co = Concrete, D = Deck, E = Earth, G = Gravel, H = Hydraulic Fill, R = Rockfill and S = Sand.

Italics: Record breaking dimensions.

Table 36. No. of over 100 m high embankment dams in the world.

Year of completion	Maximum height of dams (m)					Industrialized countries[1]
	100-149	150-199	200-249	250-300	100-300	
1930-1939	2				2	2
1940-1949	1				1	1
1950-1959	8	1			9	8
1960-1969	42	6	1		49	37
1970-1979	49	10	3		62	39
1980-1989	50	11	2	2	65	23
Totals	152	28	6	2	188	110

[1]Industrialized countries: Australia, Canada, CIS, Europe, Japan, New Zealand and USA.

also provides evidence of the fact that, starting in the 1970's, the main activity in dam construction shifted from the industrialized countries to the rest of the world, especially to Latin America and southern Asia.

6.2 GRAVITY DAMS

6.2.1 *Sound theoretical base*

Stress analysis

In his lectures published in 1826, Louis M.H. Navier (1785-1836), since 1819 professor at the school of the French 'Corps of Bridge and Highway Engineers', introduced the analysis of the stresses in structures in their elastic condition as well as the concepts of the modulus of elasticity and of safe stresses (Fig. 148) (Timoshenko 1953). Starting from formulae for the linear distribution of the stresses developed on this basis by Edouard H.T. Méry (1805-?) and professor Jean-Baptiste Belanger (1789-1874) the senior engineer of the Frech 'Corps of Bridge and Highway Engineers' J. Augustin Torterne de Sazilly (1812-1852) showed in a lecture in 1850 that the most advantageous profile for a gravity dam is a triangle with a vertical upstream face (Sazilly 1853).

Triangular cross section

In his paper, published posthumously in 1853, Sazilly also analyzed three recent French navigation dams. Their cross sections illustrate the confusion and uncertainty mentioned earlier which reigned up to that time in the design of gravity dams (Fig. 149). The Glomel and Grosbois dams even were wrongly inclined on the upstream side! The second one proved to be unstable upon impounding, so that downstream buttresses had to be added in the years from 1840 to 1842 (Guenot & Thille 1949). It was further stabilized in 1905 by a downstream impoundment and, in the 1950's, by grouting.

Figure 148. Louis M.H. Navier (1785-1836).

Sazilly's findings were first put into practice by F.X.P. Emile Delocre (1828-1908) under the supervision of M.I. Auguste Graeff (1812-1884), both members of the French 'Corps of Bridge and Highway Engineers', for the Gouffre d'Enfer flood retention dam near St. Etienne in eastern France (Fig. 150) (Delocre 1866, Graeff 1866). The 60 m high gravity dam was built from 1859 to 1866 in a narrow gorge of the Furan river (length to height ration 1.7). Notwithstanding this, it had an excessive base width of 82% of the height, because the maximum compressive stress was limited to 0.6 MPa. Moreover, it was slightly curved in plan (central angle 23°), a provision which, on Delocre's recommendation was later adopted in many modern gravity dams. Up to World War I (1914-1918) most of these were built of traditional rubble masonry, now set in Portland cement mortar (Table 37) (ICOLD 1973, 1984, INCOLD 1979, USCOLD 1988, Wegmann 1888-1927).

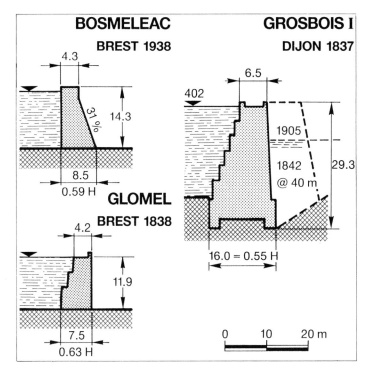

Figure 149. Cross sections of three French gravity dams of the 1830's (after Wegmann 1888–1927).

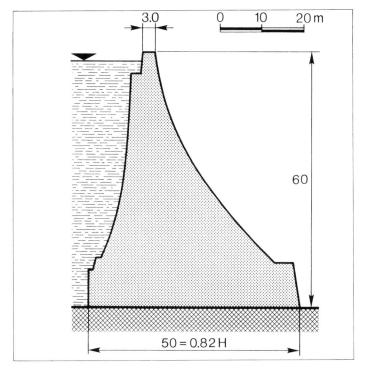

Figure 150. Cross section of the Gouffre d'Enfer dam on the Furan river, 6 km southeast of St. Etienne in eastern France (after Wegmann 1888-1927).

Table 37. World record gravity dams of rubble masonry.

Year of com-pletion	Name of dam	Country (state)	Distance in km from city	Height (m)	Length (m)	Masonry volume (1000 m^3)	Purpose
1866	Gouffre d'Enfer	France	50 SW Lyon	*60*	*102*	*40*	Flood c.
1875	Gileppe[1]	Belgium	130 E Brussels	52	*235*	*249*	Water s.
1879	Khadakvasla[2]	India	110 SE Bombay	40	*1471*	*290*	Irr./Water s.
1892	Tansa	India	Near Bombay	41	*2804*	*312*	Water s.
1902	Aswan	Egypt	700 S Cairo	39	1950	*545*	Irrig.
1904	Cheeseman	USA (Colo.)	62 SW Denver	*72*	216	79	Water s.
1906	New Croton	USA (N.Y.)	55 N New York	*91*	329	516	Water s.
1912	Aswan[3]	Egypt	700 S Cairo	44	1982	*1057*	Irrig.
1934	Mettur	India	300 SW Madras	70	1615	*1545*	Irrig./Pow.
1974	Nagarjuna	India	400 N Madras	*125*	1450	*5635*[4]	Irrig./Pow.

[1]Integrated 1970 in 15 m higher embankment dam. [2]Downstream earth backing added after impounding.
[3]After first heightening; second heightening 1934 mostly concrete. [4]Of which some 14% concrete.
Italics: Record breaking dimensions.

Uplift forces

Foundation drainage

The next larger Gileppe dam in Table 37 required the then enormous amount of 249 000 m^3 of masonry, because of its excessively conservative cross section with a base width of 132% of the height (Wegmann 1888-1927). This was due to the novel concern of the Belgian engineers about the uplift forces created by the water seeping through the dam and its foundation, an angle Sazilly and Delocre had not yet taken into consideration. However, the Belgians grossly overestimated the effect erroneously assuming that the whole dam would act as if it were submerged. A more realistic approach was followed by the designers of the 25 m high Alfeld dam, built from 1883 to 1888 some 110 km southwest of Strasbourg in then German Alsace (Fecht 1889). As is still done today, they let the uplift pressure decrease linearly from the full water pressure at the upstream face to zero at the downstream face.

Hawksley, mentioned at an earlier occasion, used a similar uplift distribution for the design of the 40 m high Vyrnwy dam, built from 1882 to 1890 some 80 km southwest of Liverpool in Great Britain (Davidson 1987-1988). The construction manager, George Deacon (1843-1909), provided transverse drains of 0.23 by 0.30 m clear section on the central half of the dam's foundation which ended in vertical shafts in the upstream third of the dam. The shafts in turn led to a longitudinal gallery above the tailwater level, which was connected by a transversal gallery to the downstream face. Deacon assumed an uplift pressure equal to the reservoir head under the upstream third of the dam and one equal to the tailwater head under the rest (Fig. 151). His pioneering drainage provisions and uplift assumptions appeared in the 1885/1886 proceedings of the Liverpool city council, whereas the design of the Alfeld dam was published in 1889.

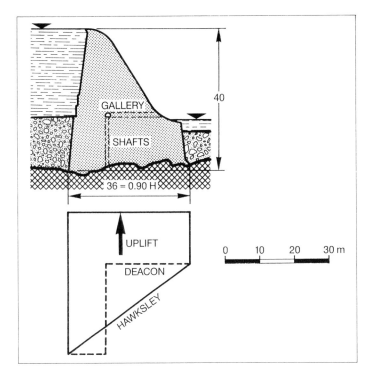

Figure 151. Cross section of the Vyrnwy dam in Great Britain and the distribution of the uplift pressures according to Deacon and Hawksley (section after Ziegler 1900-1925).

Despite these early steps in the right direction, it took more than one generation for the uplift concept – vital for the stability of a gravity dam – to be understood and accepted by the dam building community. The often quoted rule of the French engineer and scientist Maurice Lévy (1838-1910), developed after the catastrophic failure (90 deaths) of the Bouzey gravity dam 115 km southwest of Strasbourg 1895, probably brought more confusion than elucidation. The rule stipulated that on the upstream face of a dam the vertical stress due to the weight of the masonry should in any point exceed the reservoir water pressure.

Lévy's rule

The further development of the drainage provisions, e.g. by Otto Intze (1843-1904), professor at the technical university of Aachen/Germany and prolific dam designer were more promising. At the 58 m high Urft dam, built from 1901 to 1904 some 50 km southwest of Cologne, he added two rows of 0.06 m-diameter drain pipes of clay spaced 2.4 m apart behind the upstream impermeabilization with bitumen and an earthfill at the dam's heel (Wegmann 1888-1927). Similar drainage systems were adopted overseas as for example at the 56 m high Cataract dam built between 1902 and 1907 some 30 km southwest of Sydney in Australia or at the 77 m high Ashokan (or Olive Bridge) dam built from 1908 to 1914 some 140 km north of New York (Wegmann 1888-1927). Very elaborate impermeabilization and drainage pro-

Perfected drainage provisions

visions, including a grout curtain and drainage holes in the foundation rock as well as drains and galleries in the structure, were incorporated into the 107 m high Arrowrock dam built between 1911 and 1915 by the 'US Bureau of Reclamation' 23 km east of Boise in Idaho. It also happened to be the world's first large concrete dam (see below).

6.2.2 *Towards the concrete dinosaurs*

Reintroduction of
concrete

Concrete – now on the basis of Portland cement – was used for the first time since the Romans for the Boyds Corner gravity dam built from 1866 to 1872 in the USA and almost simultaneously at Pérolles in Switzerland (Table 38) (ICOLD 1973, 1984, Schnitter 1992, USCOLD 1988, Wegmann 1888-1927). The concrete was still batched by volume and contained between 250 and 270 kg of cement per m^3. At the Lower Crystal Springs dam it was mixed in cubical iron boxes revolved by steam power and transported on rails in small hand-pushed cars to the end of a trestle over the dam,

Lower Crystal Springs
dam

from where the concrete was chuted through steel tubes (Davis 1963). It was spread in layers of 0.07 m thickness and thoroughly rammed by hand. The height of the daily lifts was from 2.1 to 2.9 m and the dam was subdivided into a multitude of interlocking concrete blocks of 27 m^3 average volume (Fig. 152).

The engineer responsible for the Lower Crystal Springs dam

Table 38. World record gravity dams of concrete.

Year of comple-tion	Name of dam	Country (state)	Distance in km from city	Height (m)	Length (m)	Concrete volume (1000 m^3)	Purpose
1872	Boyds Corner	USA (N.Y.)	80 N New York	*24*	*204*	*20*	Water s.
1872	Pérolles	Switzerland	26 SW Berne	*21*[1]	195	*32*	Power
1890	Crystal Springs	USA (Calif.)	30 S San Francisco	*45*[2]	183	*120*	Water s.
1897	Periyar	India	520 SW Madras	54	439	*140*	Irrig./Pow
1915	Arrowrock	USA (Idaho)	23 E Boise	*105*[3]	351	*486*	Irrig./Flood c.
1916	Kensico	USA (N.Y.)	40 N New York	94	*562*	*738*	Water s.
1924	Schräh	Switzerland	42 SE Zürich	*112*	156	237	Power
1924	Wilson	USA (Ala.)	160 NW Birmingham	42	*1384*	*981*	Multipur.
1932	Owyhee	USA (Oreg.)	80 W Boise	*127*	254	411	Irrig.
1934	Chambon	France	130 SE Lyon	*136*	294	295	Power
1936	Hoover (Boulder)	USA (Ariz./ Nev.)	42 SE Las Vegas	*221*	379	*2486*	Multipur.
1942	Grand Coulee	USA (Wash.)	255 E Seattle	168	1272	*7450*	Multipur.
1961	Grande Dixence	Switzerland	97 E Geneva	*285*	695	5957	Power
1983	Sayano	Russia	600 SE Novosibirsk	245	1066	*9075*	Power

[1]Heightened 3 m 1909. [2]Heightened 2 m 1911. [3]Heightened 1.5 m 1937.
Italics: Record breaking dimensions.

Figure 152. Block at the Lower Crystal Springs dam, ready for concreting; note interlocking with adjoining blocks and concrete transportation trestle in background (from Davis 1963).

(Schussler, mentioned earlier) paid special attention to the control of the water content in the concrete (some 250 kg/m^3 referred to bone dry aggregates). And in fact, two years after the dam's completion, L. René Féret (1861-1957), chief of a laboratory of the French 'Corps of Bridge and Highway Engineers', published the results of tests in 1892, showing the pre-eminent influence of the ratio between the water and cement contents of a concrete on its strength (Féret 1892). The lower the ratio, the better the quality of the concrete, not only in respect to its strength, but also as to the impermeability, durability etc. The water content has, however, a minimum to ensure the workability of the concrete, whereas the cement content may need to be limited on account of the heat it develops during the hardening process. 1918 Duff A. Abrams (1880-1965) rediscovered the water/cement-ratio law for the USA and the English speaking world (Abrams 1918).

Water/cement ratio

The abovementioned heating up of the concrete during its hardening measured for the first time in 1903/1904 by electric resistance thermometres embedded in the Boonton gravity dam 40 km west of New York, also required the provision of vertical contraction joints across larger concrete dams. This was complied with for the first time at the aforementioned Ashokan (or Olive Bridge) dam and it rapidly became standard practice (Wegmann 1888-1927). Later these joints were sometimes widened, so as to act as cooling slots, subsequently to be filled with concrete only after several months. This method was first applied at the Montsalvens arch dam in Switzerland (see below). Additional cooling by circulating cold water through pipes embedded in the concrete, was introduced at the Merwin (formerly Ariel) arch dam completed 1931 some 190 km south of Seattle/Wash. (USCOLD 1988).

Heat problem

Special cements

Further effort were undertaken to reduce the heat generation at the source by developing a special low heat Portland cement first used in the Morris (formerly Pine Canyon) gravity dam, built from 1932 to 1934 some 40 km east of Los Angeles (Davis 1963). At the Bonneville dam, completed in 1938 by the 'US Army Corps of Engineers' across the Columbia River 65 km east of Portland in Oregon, 25% of the Portland cement were replaced for the first time by a calcined pozzolan (Davis 1963). In conjunction with cement or lime (see Roman concrete), it develops strength but no heat. Finally precooling of the concrete was introduced either by chilling the aggregates and the mixing water or by replacing the latter with chipped ice, which absorbs a considerable amount of heat in the melting phase. The latter method was used for the first time by the 'US Army Corps of Engineers' at the Norfork flood

Figure 153. Chutes for distribution of the concrete at the Schräh dam built between 1922 and 1924 in Switzerland, which later experienced extensive frost damage (photo E. Schnitter, Itschnach/ZH).

control and power dam, completed in 1944 about 170 km north of Little Rock in Arkansas.

A rather unfortunate innovation had been introduced by the 'US Bureau of Reclamation' in 1912 at the abovementioned Arrowrock dam (USCOLD 1988). By increasing the water content of the mix in order for it to be more easily chuted into the forms, it proved in many instances to result in an insufficiently durable concrete, especially under severe frost (Fig. 153). At the abovementioned Morris dam, a method to compact the concrete by means of immersed vibrators was introduced (Davis 1963). It permitted a return to much lower water contents, in keeping with Féret's and Abrams' laws. A further improvement of concrete durability was to be achieved later by the addition of an air entraining agent to the concrete mix, as was done for the first time at the Angostura gravity dam, completed in 1949 by the 'US Bureau of Reclamation' 275 km southwest of Pierre in South Dakota.

Figure 154. John L. Savage (1879-1967).

In the meantime, the 'US Bureau of Reclamation', from 1924 under its chief design-engineer John L. Savage (1879-1967) (Fig. 154), had proceeded to a quantum jump at the Hoover (formerly Boulder) dam. Not only was it 60% higher and 2.5 times larger than any existing dam (Table 38), but it was also to impound the unheard of quantity of 38 550 million m^3 of water (USCOLD 1988). In understandable deference to such a feat the dam, built between 1931 and 1936, was designed as a massive curved gravity dam, although it had been contemplated as an arch dam and much pioneering effort had been invested in the study of the behavior of arch dams (see below). Concreting of the Hoover dam barely took two years with some 5000 men placing up to 8000 m^3 of concrete per day! It was compacted with vibrators and cooled by circulating water through embedded pipes (Fig. 155).

Hoover dam

Figure 155. Downstream view of 221 m high Hoover dam built from 1931 to 1936 on the Colorado river in the southwestern USA (photo by the author).

More concrete
'colossuses'

Twice such concrete placement rates were required at the Grand Coulee straight gravity dam. Also built by the 'US Bureau of Reclamation' in the period from 1933 to 1942, its volume was three times as large as that of Hoover dam (Table 38) (USCOLD 1988). Thereafter, several more, though smaller 'colossuses' were built in the USA and other countries. The tallest of them still is the 285 m high Grande Dixence dam constructed from 1951 to 1962 high up in the Swiss Alps (Table 38) (Fig. 156). Impounding a mere 400 million m^3 of water, it nevertheless governs a power head of 1882 m. However, since reaching their prime in the 1960's, gravity dams have become something like a dying species, the dinosaurs of an era which especially the industrialized countries with their high labour cost no longer can afford (Fig. 145, Table 39).

Figure 156. 285 m high Grande Dixence gravity dam in Switzerland on a stamp issued in 1985 on the occasion of the 15th Int. Congress on Large Dams held in Lausanne.

Table 39. No. of over 100 m high gravity dams in the world.

Year of completion	Maximum height of dams (m)					Industrialized countries[1]
	100-149	150-199	200-249	250-300	100-300	
1910-1919	1				1	1
1920-1929	4				4	4
1930-1939	8		1		9	8
1940-1949	4	2			6	5
1950-1959	16	1			17	12
1960-1969	22	2	1	1	26	20
1970-1979	14		2		16	11
1980-1989	13	4	1		18	10
Totals	82	9	5	1	97	71

[1]Industrialized countries: Australia, Canada, CIS, Europe, Japan, New Zealand and USA.

6.2.3 *Roller compacted concrete revival?*

For all that, due to the accelerated assimilation of concrete placement techniques to the highly mechanized embankment construction (see above), the above obituary for the gravity dam might
be premature. Bulldozers were introduced in Switzerland already
in the early 1950's for the spreading of the concrete whereas vibrators found themselves attached to caterpillar tractors (Fig. 157). For
the 174 m high Alpe Gera gravity dam, completed in 1964 some
110 km northeast of Milan in Italy, rear-dump trucks were used to
distribute the concrete on the dam instead of the usual buckets

Mechanization of
concreting

Figure 157. Mechanization of
concrete distribution (above)
and vibration (below) on Swiss
dams (photos R. Hediger and E.
Schnitter).

suspended from cableways spanning the valley (Fig. 158) (ENEL 1970). However, this implied that the dam could no longer be built up by independent blocks and that the contraction joints had to be cut into the concrete after placement. Moreover, the latter's cement content was reduced to as little as 115 kg/m^3 whilst including 50% of blast furnace slag which acts like a pozzolan.

From 1978 to 1981 a similar, though smaller gravity dam was built by the 'Japanese Ministry of Construction' at Shimajigawa, 350 km west of Osaka, in which vibratory rollers were used to compact the concrete instead of the tractor-mounted immersed vibrators employed at the Alpe Gera dam (Fig. 159) (MOC). The joints as well as most forms were eliminated at the 52 m high and 543 m long Willow Creek flood control dam built between 1981

Roller compacted concrete

Figure 158. Upstream view of Alpe Gera gravity dam built between 1961 and 1964 in northern Italy; at upper left concrete mixing plant above cable cars, lowering the concrete to a retractable loading station for rear-dump trucks (from ENEL 1970).

Figure 159. Construction scene at the Shimajigawa gravity dam in western Japan; on right background: vibratory roller and, right: parked joint cutting machine (photo MOC undat.).

Table 40. Roller compacted concrete dams built (and under construction) in the 1980's (WPDCH).

| Country | Maximum height of dams (m) | | | |
	15-49	50-99	100-150	15-150
Australia	3 (0)			3 (0)
China	3 (0)	4 (5)	0 (2)	7 (7)
Japan	1 (1)	5 (5)	1 (6)	7 (12)
South Africa	3 (2)	2 (0)		5 (2)
Spain	3 (1)	1 (2)		4 (3)
USA	8 (1)	4 (0)		12 (1)
Various	4 (2)	3 (4)	0 (1)	7 (7)
Totals	25 (7)	19 (16)	1 (9)	45 (32)

and 1983 by the 'US Army Corps of Engineers' about 250 km east of Portland in Oregon (Schrader 1982). In addition, the content of cement, which included 29% of fly ash as a pozzolanic material, was reduced to 66 kg/m^3 of concrete.

Throughout the world some 45 roller compacted concrete dams of over 15 m in height were completed and 32 started in the 1980's, whereby the height of the latter increased significantly (Table 40). Due to simplifications and acceleration of the speed of their construction, gravity dams could now compete successfully with rockfill embankments (see above), not least because spillways and outlets could be incorporated with relative ease (Dunstan 1990). For the most recent structures, the trend points to an increase of the cement content of the concrete over the abovequoted figures with a view of improving its impermeability and durability. At the same time, the portion of pozzolanic material in the cement had to be augmented in order to minimize the development of heat during hardening.

Rapid spread of new technique

6.3 BUTTRESS DAMS

Today we understand buttress dams as derivations from the massive gravity type with the introduction of intermediate spaces (Fig. 160). These spaces allow water seepage through foundation and dam to discharge not only downstream, but also side- and upwards into them, thus greatly reducing the uplift pressures. For the classical triangular profile with a vertical upstream face, the total elimination of uplift pressures would theoretically have permitted a reduction of the dam's mass by 40% without reducing its stability. However, the actual reduction proved considerably smaller, because the intermediate spaces had to be closed upstream with slabs, arches or a thickening of the buttress heads to make them contiguous. The saving in costs was even more modest due to the fact that

Derivation of buttress dams

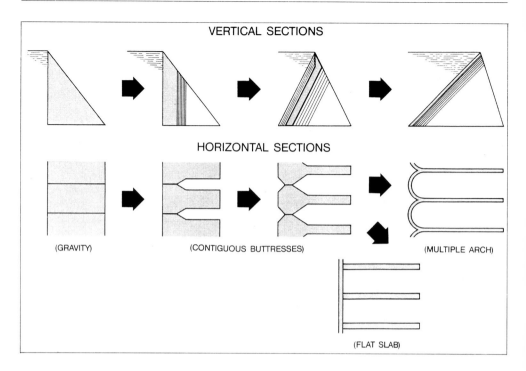

VERTICAL SECTIONS

HORIZONTAL SECTIONS

(GRAVITY)   (CONTIGUOUS BUTTRESSES)   (MULTIPLE ARCH)

(FLAT SLAB)

Figure 160. Derivation of the buttress dam types from the massive gravity dam.

buttress dams, per m^3 of concrete, required more formwork of a more complicated shape and, often, a higher cement content in the less easily placed concrete. On the other had the intermediate spaces facilitated the dissipation of the heat produced during the hardening of the concrete, so that elaborate cooling measures were seldom needed during the construction of a buttress dam.

Given the absence of uplift, more substantial savings were possible by also inclining the upstream face and thus mobilizing the vertical water load on it for sliding stability. This knowledge had already been gained by Italian and Spanish engineers in the 16th century (see above). If the stability, in terms of the inclination of the resultant of the forces, was maintained the same as with the vertical upstream face, a buttress dam inclined 50% on both faces permitted a theoretical reduction of the dam's mass by 64%. At 100% upstream and 33% downstream inclinations the reduction was as much as 89%! Also the eccentricity of the resultant was greatly improved, since it moved from the downstream third of the section in the case of the vertical upstream face to almost its centre with the 100% upstream inclination. Correspondingly, the vertical stresses, were almost uniform over the section, although five times larger. Except for high dams, this was welcome because it permitted a better utilization of the strength of the construction material.

Inclination of upstream face

6.3.1 *Prime of multiple arches*

As already mentioned, it took almost 100 years until the ingenious feat of a British 'Royal Engineers' member with his Mir Alam multiple arch dam near Hyderabad in India in 1804 was repeated on a more modest scale in Australia between 1896 and 1897. At Junction Point Reefs, 210 km west of Sidney, one Oscar Schulze built a 19 m high dam consisting of five arches with spans of 8.5 m across the Belubula river (Fig. 161) (Schulze 1897). These were inclined upstream and of elliptical shape. Moreover, the arches, built of brick like the buttresses, were covered with concrete on the upstream face, so that it became a straight inclined plane, like in Villareal's 18th century dams in Spain (see above). Schulze analyzed his structure as a bridge turned through 60°.

Belubula dam

Was it pure coincidence that, while the Belubula dam was under construction, a similar design was being studied for a 31 m high structure near Ogden, 56 km north of Salt Lake City in Utah (Goldmark 1897), or, was there some kind of transpacific technology transfer as in the case of single-span arch dams (see below)? At that time both regions, Australia and the West of the USA were undergoing similar pioneering developments favouring solutions with a minimum of imports and transports.

Ogden project and steel dams

One alternative for the upper half of the Ogden dam that was never built even envisaged a rather light steel structure of concave (downstream curved) sheets resting on trusses or struts spaced 3 m apart and braced together. Such or similar designs were realized in 1898 at the 14 m high Steel (or Ashfork) dam 70 km west of Flagstaff in Arizona, 1901 at the 23 m high Redridge dam 20 km

Figure 161. Partial downstream elevation and cross section of Belubula multiple arch dam in Australia (after Schulze 1897).

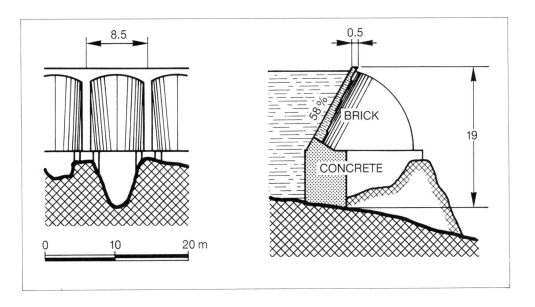

Figure 162. Recent down-stream view of the disused Red-ridge steel dam in Michigan; the dam's foundation was also used for a railroad trestle (photo T.S. Reynolds, Houghton/MI).

north of Houghton in Michigan and 1907 at the 25 m high Hauser Lake dam 24 km north of Helena in Montana (Fig. 162) (Reynolds 1989). The central portion of the Hauser Lake dam rested on gravel which washed out in 1908, despite an 11 m deep cut-off of sheetpiles, entailing the collapse of the steel structure. It was replaced from 1909 to 1912 by a concrete gravity dam founded throughout on rock.

In the meantime, the first multiple-arch dam of reinforced concrete had been completed in 1908 in only 114 (!) working days. It impounded the Hume Lake fluming reservoir on the Ten Mile creek in the Californian Sierra Nevada mountains. The dam was designed and supervised during its construction by direct labour, by John S. Eastwood (1857-1924). Immediately after his graduation from the University of Minnesota as a civil engineer he went

Reinforced concrete multiple-arch dams

to work in the West of the USA and became chief engineer for a high-head power plant on the San Joaquin river 60 km north of Fresno (Jackson 1979). Afterwards he worked on the design of the first phase of the later famous Big Creek project in the same area, until he set up his own consulting practice in 1910.

The 20 m high Hume Lake dam comprised twelve circular arches of 15.2 m span and with a central angle at the crest of 124° (Fig. 163) (USCOLD 1988). In the upper 5 m of the dam, the arches were vertical and some of them had spilling openings, whereas, below they were inclined 63% upstream. The arches as

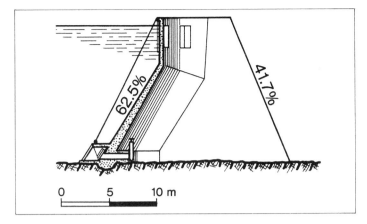

well as the buttresses were very thin, so that the whole dam required just 1700 m³ of concrete or barely 10% of an equivalent gravity structure. The reinforcement consisted of old logging cables and railroad scrap iron. Placed mainly in the arches, it served principally to resist tensile temperature stresses, while those produced by the water load were almost exclusively compressive. For the analysis of the arches, Eastwood used the simple 'cylinder formula', according to which the arch stress equals pressure times radius divided by thickness (see below). He designed the buttresses for zero eccentricity of the resultant of the forces.

Hume Lake dam

After completion of the Hume Lake dam, Eastwood was entrusted from 1910 to 1911 with the replacement of the extremely slender Big Bear Valley single-arch dam built 27 years earlier by Frank E. Brown (1856-1914) in the San Bernardino mountains 130 km east of Los Angeles (see below). The new, 28 m high dam was made some 9 m higher than the old one. As a result it could store the threefold amount of water for irrigation (Eastwood 1913). Interestingly enough, the total costs of both dams were about equal, and this inspite an inflationary factor of some 20% that had meanwhile occurred. This clearly demonstrated, once more, the economy of Eastwood's new design. It was modified in the Big Bear Valley dam only by the addition of heavily reinforced, horizontal strut-ties between the buttresses near the crest as well as at two and one thirds of their height. This provision increased their lateral stability in case of 'seismic disturbances or vibrations of any kind' (Eastwood 1913). It was increased further in 1988 by quadrupling the dam's mass (Denning 1993).

Second Big Bear Valley dam

Overall, a dozen multiple-arch dams were built in the following decade and a half according to Eastwood's design, among these, in 1918, the record breaking, 41 m high Hodges dam 37 km north of San Diego in California (Table 41) (ANIDEL 1961, ICOLD 1973, 1984, Noetzli 1924, USCOLD 1988, Wegmann 1888-

Table 41. World record multiple-arch dams of concrete.

Year of completion	Name of dam	Country (state)	Distance in km from city	Height (m)	Arch span (m)	Inclination % US[1]	DS[1]
1908	Hume (Ten Mile)	USA (Calif.)	320 SE San Francisco	20	15.2	63	42
1911	Big Bear Valley	USA (Calif.)	130 E Los Angeles	28	9.8	75	50
1916	Gem	USA (Calif.)	280 E San Francisco	34	12.2	84	10
1918	Hodges	USA (Calif.)	37 N San Diego	41	7.3	100	25
1924	S. Chiara (Tirso)	Italy (Sard.)	90 N Cagliari	70	15	65	35
1927	Pleasant[2]	USA (Ariz.)	40 N Phoenix	78	18.3	100	15
1928	Coolidge	USA (Ariz.)	140 E Phoenix	76	54.9	60[3]	30[3]
1939	Bartlett	USA (Ariz.)	60 NE Phoenix	87	18.3	90	36
1959	Grandval	France	160 SW Lyon	88	50	70	40
1968	D. Johnson	Canada	470 NE Quebec	214	76/162	60	60/65

[1]US = Upstream face, DS = Downstream face (buttress). [2]Replaced 1992 by an embankment dam. [3]Average.

Figure 164. Fred A. Noetzli (1887-1933).

Double-curved
dome-structure

1927). The concept was quickly adopted by such other renowned dam designers in the western USA as Lars J. Jorgensen (1876-1938), better known for his pioneering single-arch design (see below), and by Fred A. Noetzli (1887-1933) (Fig. 164). Noetzli a Swiss immigrant, Sc.D. in surveying attained from the Federal Institute of Technology in Zürich, in 1924 proposed double-walled, hollow buttresses with interior diaphragm bracings to give them adequate lateral stability for still greater dam heights and arch spans (Noetzli 1924). This improvement was first applied to the 78 m high Lake Pleasant (or Waddell) dam completed some 40 km north of Phoenix in Arizona with dubious success, since extensive reinforcements became necessary after the first filling of the reservoir (USCOLD 1975, 1988). The application of the double-walled buttress was wholly successful a decade later at the 87 m high Bartlett dam 60 km northeast of Phoenix, built from 1936 to 1939 by the 'US Bureau of Reclamation' (USCOLD 1988). Despite their record heights, however, both dams had relatively modest arch spans of 18.3 m.

As mentioned before, a decidedly larger arch span of 54.9 m was achieved in 1928 at the 76 m high Coolidge dam of the 'US Bureau of Indian Affairs' 140 km east of Phoenix (USCOLD 1988). With a thickness of only 1.2 to 6.1 m, the three arches were also curved vertically. Apart from the single-span arch dam designed for the Ithaca, New York water supply by Gardner S. Williams (1866-1931), professor of hydraulic engineering at the Cornell University in the same city, this was the first application of a double-curved dome-structure in dam engineering. Construction of the Ithaca dam had been discontinued 1903 at one third of the planned height of 27 m (Williams 1904).

After World War II (1939-1945) large-span multiple-arch dams

Figure 165. Downstream night view of the 214 m high D. Johnson multiple-arch dam in northern Canada with a central arch span of 162 m (photo Hydro Quebec, Montreal/ Canada).

were pursued further by the prominent French dam designer André Coyne (1891-1960). In pre-war Europe, multiple-arch structures had attained a certain importance only in Italy during the 1920's (ANIDEL 1961). After the war, several of them were built in France, among which some sizable structures. However, Coyne's most important project of this type and the apogee of the multiple-arch dams, was built in the years from 1961 to 1968 in northern Canada. On the Manicouagan river 470 km northeast of Quebec it formed the D. Johnson reservoir of 141 900 million m^3 capacity (Fig. 165). Unfortunately, the elegant structure was plagued by numerous progressive cracks (Bulora et al. 1991).

Further multiple-arch dams

6.3.2 *Heyday of flat-slab buttress dams*

In the wake of enthusiasm for the still relatively new but fast spreading use of reinforced concrete – which at the beginning of the 20th century also fascinated quite a few dam builders – the hottest competitor to the multiple-arch dam was the flat-slab buttress type developed and patented by Nils F. Ambursen. Born 1876 in Norway, he emigrated to the USA immediately following his graduation from the Porsgrund engineering college as mechanical engineer at the age of 18 (!). Barely five years later he set up the firm of 'Ambursen & Sayles' in Watertown in northern New York State, which in 1903, became the 'Ambursen Hydraulic Construction Co.' of Boston in Massachussetts.

In the same year, the new firm completed its first rather modest dam of the novel design at Theresa, 30 km north of Watertown, within a mere 18 (!) working days (Ambursen & Sayles 1903). Interestingly enough, its main competitor was not a gravity dam, as in the case of the multiple-arch dams, but the traditional timber

Ambursen's flat-slab buttress dam

Theresa dam

dam described earlier (Fig 133). Like its log type competition, Ambursen's design took full advantage of the stabilizing effect of the vertical water load on the strongly inclined upstream face, which required a minimal buttress thickness per unit of dam length (Fig. 166). This corresponded well with the rather narrow buttress spacing dictated by the limits on the bending stresses in the upstream slab. The latter was articulated by contraction joints along its interfaces with the buttresses supporting the slab on protruding haunches. From the second, three times higher project at Schuylerville 50 km north of Albany in New York State onwards, the lateral stability of the buttresses was assured by horizontal brace beams between them, as in multiple-arch dams.

Popularity of flat-slab buttress dams

The Ambursen dam type soon became quite popular and by the end of 1910 almost 60 such dams had been built (Table 42) (Ambursen & Sayles 1903, ANIDEL 1961, Rec. Hydr. 1976, Tolle et al. 1979, Wegmann 1888-1927). Among them appears the record-breaking 41 m high La Prele (or Douglas) irrigation dam,

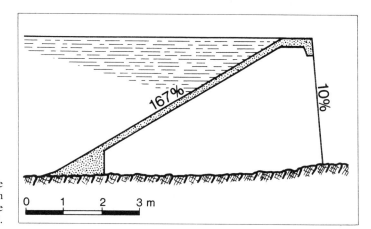

Figure 166. Cross section of the Ambursen flat-slab buttress dam at Theresa in New York State (after Ambursen & Sayles 1903).

Table 42. World record flat-slab buttress dams.

Year of completion	Name of dam	Country (state)	Distance in km from city	Height (m)	Slab span (m)	Inclination % US[1]	DS[1]
1903	Theresa	USA (N.Y.)	410 NW New York	3	1.8	167	10
1904	Schuylerville	USA (N.Y.)	250 N New York	9	2.5	100	65
1906	Ricketts	USA (Pa.)	160 W New York	16	?	?	?
1907	Ellsworth	USA (Maine)	320 NE Boston	20	4.6	100	58
1909	Prele (Douglas)	USA (Wyo.)	190 N Cheyenne	41	5.5	119	9
1916	Combamala	Italy	70 SW Turin	42	5.5	70	29
1935	Lampasas	USA (Tex.)	250 NW Houston	51	?	?	?
1937	Rodriguez	Mexico	10 SE Tijuana	73	6.7	100	20
1948	Escaba	Argentina	1000 NW B. Aires	83	12.2	100	27

[1]US = Upstream face, DS = Downstream face (buttress).

built from 1908 to 1909 in eastern Wyoming (Tolle et al. 1979). It required 17 200 m^3 of concrete or 43% of the amount needed for an equivalent gravity dam. The average reinforcement content was 23 kg/m^3 of concrete. The dam had to be provided with a new upstream slab in 1977-1979 since the original one had disintegrated by about 20% of its thickness in the rough mountain climate. At the same time, underseepage was reduced to approximately one-third or 0.28 m^3/s (still substantial!) by grouting below the upstream cut-off wall and into the alluvium left in place below the foundation slab across the valley floor.

La Prele dam

The provision of a foundation slab seemed to defeat the very principle on which the stability of buttress dams rested. The slab's foundation therefore had to be thoroughly and permanently drained. Quite a few Ambursen dams were built on alluvium, a circumstance which was even interpreted as representing one of their additional advantages, after the 13 m high Ashley (or Pittsfield) dam 170 km west of Boston in Massachussetts survived the wash-out of its alluvial foundation undamaged in 1909. Less lucky, however, were a 5 m high structure at Dansville in New York State, a 3 m high one near Janesville in Wisconsin and the 16 m high Stony dam in West Virginia. They all failed, 1909, 1912 and 1914 respectively, following piping in their alluvial foundations (USCOLD 1975, 1988).

Buttress dams with foundation slab

Nevertheless, Ambursen-type dams continued to proliferate and, by the end of the 1920's, more than 200 had been constructed, thus outnumbering multiple-arch dams by far. Outside the USA, the flat-slab buttress design was especially adopted after World War II (1939-1945) for about 50 dams exceeding 15 m in height in Ambursen's native Norway (Berdal 1968). This was primarily due to the fact that many sites there were difficult to reach with trans-

Norwegian flat-slab buttress dams

Figure 167. Downstream view of the Combamala flat-slab buttress dam built from 1915 to 1916 in northwestern Italy (photo Motor-Columbus, Baden/ Switzerland).

Other flat-slab buttress dams

ports of construction materials and because structures of only moderate height were required. To improve the dams' resistance to the severe winter climate, the Norwegians usually added a thin, vertical insulating wall between the downstream part of the buttresses. Moreover, they reduced the number of contraction joints in the upstream slab by making it continuous over two or three buttresses. Almost no flat-slab buttress dams were built in the rest of Europe and only a few in Latin America (Fig. 167). There the highest of them all was completed in 1948 at Escaba in the northwestern Argentine province of Tucuman reaching 83 m (Table 42).

6.3.3 *Survival of contiguous buttresses*

Demise of reinforced concrete dams

Despite the temporal popularity of reinforced concrete dams, especially of the Ambursen flat-slab type, the rise of labour costs in relation to the prices of materials and the ensuing mechanization, worked strongly against them. Moreover, several of them deteriorated at an unexpectedly fast pace under the severe climatic conditions to which they were sometimes exposed, or they otherwise developed cracks and leaks. The wisdom of using thin concrete members in dams at all, was being increasingly questioned. Whereas the share of buttress dams in the more than 15 m high category had already started to drop in the USA after World War I (1914-1918), this phenomena took place in Western Europe only after World War II (1939-1945) (Fig. 168).

Figure 168. Development of the share of buttress dams in all dams of more than 15 m height built in the USA and Western Europe.

The latter development was mainly due to the introduction of a new kind of design in which the arches or slabs as upstream

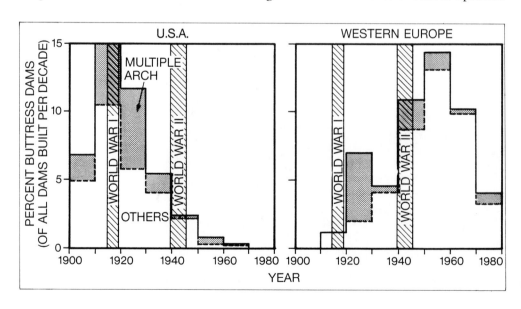

closures of the intermediate spaces between the buttresses were replaced by a thickening of their heads to make them contiguous. The thicker the buttresses and their massive heads, the more amenable they became to modern mass concrete techniques. Such a solution was used 1927-1928 by Noetzli (see above) for the 35 m high spillway section of the V. Carranza (formerly Don Martin) irrigation dam in northeastern Mexico (Noetzli 1928). The same is true for the 26 m high Burgomillodo dam, completed in 1929 some 100 km north of Madrid by the Spanish engineer Federico Cantero (1872-1946) (Gomez-Navarro & Aracil 1958). While the latter's upstream face was inclined only 21%, Noetzli on average adopted a 46% inclination to take greater advantage of the stabilizing effect of the vertical water load and to get thus as wide a spacing of the buttresses as the bending stresses in their heads would permit. To reduce the latter, he gave the upstream face of the heads a convex shape, from where the name of the new design 'round-head buttress' originated (top left of Fig. 169). Similarly, the polygonal approximation of the round head later became known as the 'diamond-head buttress'. Moreover, Noetzli introduced T-

Development of contiguous buttress dams

Noetzli's design

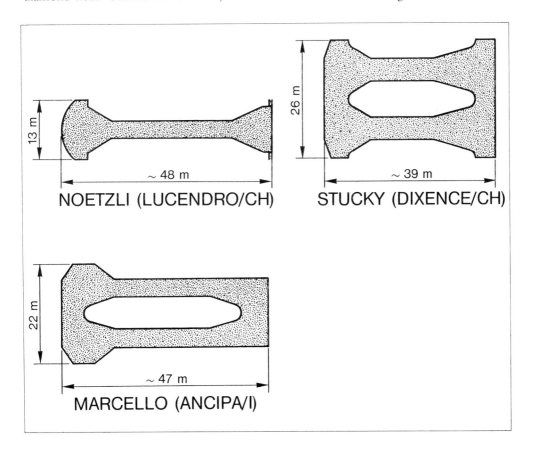

Figure 169. Comparative horizontal sections of contiguous buttress dams 50 m below their storage water levels.

Figure 170. Claudio Marcello
(1901-1969).

Marcello's design

shaped enlargements at the downstream ends of the buttresses for a better distribution of the stresses, which are highest there, and to reduce the clear span of overflow or insulating decks.

In deference to the high altitude of about 2200 m a.s.l. the downstream T's were made contiguous at the Dixence power dam in southwestern Switzerland (Stucky 1946). Moreover, contraction joints were provided for at only every second intermediate space, resulting in II-shaped buttresses (top right of Fig. 169). The upstream face of Dixence dam was almost vertical, but, in plan, it was curved for additional stability. The maximum height of 87 m was a world record for buttress dams at the time of its completion in 1935, record that was held until after World War II (1939-1945). The dam was designed by Alfred Stucky (1892-1969), professor for applied hydraulics at the Institute of Technology in Lausanne, Switzerland. It was submerged in 1957 in the reservoir of the 285 m high Grande Dixence gravity dam (see above).

The Dixence design was adopted during World War II (1939-1945) by Claudio Marcello (1901-1969) (Fig. 170), chief engineer for hydropower plants at the Edison company of Milan in Italy, for the 54 m high Trona and the 41 m high Inferno power dams 70 km north of Milan (Marcello 1955). He omitted the downstream T's in the intermediate spaces having contraction joints. For the numerous buttress dams designed thereafter by Marcello, he foresaw an inclination of 45-50% on either face and carefully standardized both, the shapes for buttress heads and the intermediate spaces (bottom of Fig. 169).

In fact, for most of the buttress dams built in the last decades, either Marcello's double-buttress or Noetzli's single-buttress designs were adopted, sometimes with rather trivial variations (Table 43) (ANIDEL 1961, ICOLD 1973, 1984, Noetzli 1928, Stucky

Table 43. World record contiguous buttress dams.

Year of completion	Name of dam	Country (state)	Distance in km from city	Height (m)	Buttress spacing (m)	Inclination % US[1]	DS[1]
1928	Carranza (Martin)	Mexico	200 N Monterrey	35	9	46[2]	60[2]
1935	Dixence	Switzerland	80 SE Lausanne	87	13	4	78
1950	S. Giacomo Fraele	Italy	140 NE Milan	96	15	3	62
1953	Ancipa	Italy (Sicily)	110 E Palermo	105	11	45	45
1959	Albigna	Switzerland	140 SE Zürich	115	20[3]	0	66[2]
1962	Hatanagi I	Japan	140 W Tokyo	125	11/16	50	50
1969	Oriol (Alcantara)	Spain	280 W Madrid	135	11	45	45
1983	Itaipu	Brazil/Paraguay	310 E Asunción	196	17	58	46

[1]US = Upstream face, DS = Downstream face (buttress). [2]Average. [3]Buttress thickness 75 % of spacing (against the usual 35-40%).

1946). In a general way, however, both dam types became relatively less important for the same economic reasons that prior to World War II (1939-1945) had already caused the multiple-arch and flat-slab buttress dams to become obsolete. Nevertheless, the largest hydropower plant ever built so far, namely the 12 600 MW Itaipu scheme on the Brazilian-Paraguayan border has a double-buttress main dam with an unprecedented height of 196 m and a long wingwall of single buttresses (Fig. 171).

Figure 171. Downstream aerial view of the Itaipu dam, power plant and spillway on the border between Brazil and Paraguay (photo Siemens/KWU, Erlangen/D).

6.4 ARCH DAMS

6.4.1 *The pioneers*

By arch dams, we interpret masonry or concrete structures, the base width of which measured less than half their height and which thus had to rely on their curvature in plan for the transmission of part of the water load laterally into the valley flanks. Invented by the Romans, this dam type experienced a short florescence in Iran under the Mongolians and sporadically emerged in the 17th century in Spain and the Alpine countries (see above). In 1824-1825 the previously mentioned Pontalto flood control dam in the (at that time Austrian) South Tyrol, was heightened by 8 m to 25 m total height (Noetzli 1932). It was probably known at first hand by

Definition of arch dams

Figure 172. François Zola (1795-1847) (from a painting in the Musée Emile Zola, Médan/ France).

François Zola (1795-1847) (Fig. 172) – the father of the famous French writer Emile Zola (1840-1902) – for he was born and trained as surveyor in nearby (at that time also Austrian) Venice and in the early 1820's worked on the construction of a railway near Linz in northeastern Austria (Rigaud 1957).

In 1837, five years after he had established himself as a consulting engineer in Marseilles in southeastern France, Zola took part in a competition for a new water-supply scheme to nearby Aix-en-Provence. He proposed an aqueduct from the Infernet creek east of the city, whereby the large seasonal variations of the relatively modest runoff were to be regulated by three consecutive reservoirs. These were all to be formed by arch dams. After a long fight for his ideas, Zola was finally able to initiate the work on the first of the dams in 1847, but he died just two months later. Deprived of its leader, the work progressed only haltingly, to be completed but seven years later in 1854.

Named after its designer (a rare thing indeed!), the Zola dam was 43 m high, modestly curved to a central angle of 77° and fairly sturdy (Fig. 173). Besides being a very fine structure, still in excellent shape to-day, it was also the first arch dam designed on the basis

Figure 173. View along the Zola arch dam completed in 1854 some 5 km east of Aix-en-Provence in southeastern France (photo by the author).

of an analysis of the stresses in it. Zola's calculation supposed that the dam consisted of a stack of independent arches, each of which having to resist the total pressure corresponding to its depth below the storage water level. The average stress in every arch was determined by the 'cylinder formula' (the product of pressure and radius divided by the arch thickness). This simple relationship had already been experimentally established by the French physicist Edmé Mariotte (1620-1684). It was formalized in 1826 by the above mentioned Navier, who also laid the foundations for the theory of elastic arches (Timoshenko 1953). The latter theory was formulated only in 1854 by Jacques A.C. Bresse (1822-1883), who like Navier was a French teacher and researcher in civil engineering sciences. Notwithstanding the simple calculation Zola thus had to apply, the stresses in his dam were satisfactory even by analysis according to the modern crown-cantilever method: only compressive vertical stresses of up to 0.7 MPa; up to 0.9 MPa horizontal compression and 0.4 MPa horizontal tension (without thermal effects).

Despite this pioneering example, almost 80 years went by before the next French arch dam of similar size was built on the Bromme river in the Central Massif. This may have resulted from the fact that the earliest description of the Zola dam appeared as late as in 1872 and, this furthermore in another context (Tournadre 1872). It met with a wider and well deserved publicity in the first standard work on 'The Design and Construction of Dams', published in 1888 by the Swiss-American hydraulic engineer Edward Wegmann (1850-1935) (Wegmann 1888-1927). In the above mentioned basic papers of 1866 by Delocre and Graeff it had furthermore been passed over, although their authors described all the old Spanish dams and outlined a method of analysis for arched structures very similar to that used by Zola (Delcore 1866, Graeff 1866). In a study of 1879, G. Albert Pelletreau (1843-1900) found that in order to minimize the volume of an arch dam, its radius of curvature should decrease from crest to base. This, according to the 'cylinder formula' also yields smaller arch thicknesses for a given allowable stress (Pelletreau 1879).

Almost in parallel to the Zola dam, a small arch dam was built 1855/1856 near Parramatta east of Sydney in Australia (Aird 1961). It was designed by Captain Percy Simpson (1787-1877), who had come to Australia in 1822 as commander of a convict settlement and who later became surveyor and land commissioner. The dam showed a considerable length-to-height ration and was remarkably slender (Fig. 174). In 1898, under the direction of Cecil W. Darley (1842-1928) about 3 m of concrete was added to the top of the masonry structure. Darley emigrated from Great Britain in 1867 and had become chief engineer of the 'Public

Zola arch dam

Analysis of arches

Zola dam overlooked by profession

Parramatta thin arch dam

Figure 174. Crest of the Parramatta arch dam completed in 1856 and heightened in 1898 east of Sidney in Australia (photo M. Dreier, Baden/AG).

Works Department of New South Wales'. Concurrently he initiated the construction of a whole series of very thin arch dams in the Blue Mountains behind Sydney that were to provide small water-supply reservoirs in the newly settled arid areas (Table 44) (Nimmo 1966, Wade 1908-1909).

More thin arch dams in Australia

These Australian structures all had one vertical cylindrical face. They were up to 36 m high and built entirely of concrete, which was batched by volume, hand mixed, transported in wheel barrows, placed in thin layers and compacted by tamping. It contained roughly 280 kg/m^3 of cement, but reached only about one quarter of the strenght attainable today with a similar mix. However it was sufficient to cover the average stresses of up to 2.6 MPa calculated with the 'cylinder formula' for independent arches, although only a narrow safety margin was left with respect to local stress concen-

Table 44. Early thin arch dams in Australia (over 15 m high).

Year of completion	Name of dam	Distance from Sydney (km)	Height (m)	Length[1]/ height	Base width/ height	Crest slenderness[2]	Central angle °
1856/1898	Parramatta	21 NW	16	4.3	0.29	33	80
1898	Tamworth	325 N	20	6.8	0.33	83	100
1899	Mudgee	210 NW	15	10.0	0.22	84	117
1899	Wellington	260 NW	15	7.0	0.20	50	133
1903	Barossa	1100 W	36	4.0	0.29	45	136
1906/1915	Lithgow 2	112 NW	27	2.5	0.27	29	126
1907	Medlow	93 NW	20	1.5	0.14	17	78

[1]Arch only (without abutment blocks and/or wing walls). [2]Ratio of radius of curvature to thickness.

trations resulting from the boldness of the dams' design. The most daring was the 15 m high Mudgee dam which, notwithstanding its extreme length-to-height ration of 10, was very thin.

When Darley started the construction of his arch dams he could rely not only on the example set by the Parramatta dam, but also on that of the 20 m high Abbeystead dam, completed in 1881 for the water supply of Lancaster, 60 km northwest of Manchester (Mansergh 1881-1882). The sturdy structure was strongly curved (central angle 120°) and was the first arch dam to be built entirely of concrete mechanically mixed as done later at the Crystal Springs gravity dam (see above). Moreover, Darley was aware of the extremely bold Big Bear Valley irrigation dam, completed in 1884 in the San Bernardino mountains about 130 km east of Los Angeles in California to the design of the young engineer and land developer Frank E. Brown (1856-1914). He may in turn have been inspired by the only 9 m high Stone dam built in 1871, with similar structural ratios as the Parramatta dam, 30 km south of San Francisco by the aforementioned Schussler. While only moderately curved to a central angle of 46°, Brown's thin dam set a record with its crest slenderness of 105. Its masonry was laid from rafts floating on the rising reservoir. Although it performed satisfactorily, it was replaced in 1911 by a higher multiple-arch dam designed by Eastwood (see above).

Southeast of San Diego in California, the similar Upper Otay dam built from 1898-1901 with masonry and concrete, is still in operation, but its storage water level was reduced in 1985 (US-COLD 1988). Slightly north of it, the Sweetwater irrigation dam had also been started by Brown in 1886 as a thin masonry arch, but was changed to a far sturdier structure when James D. Schuyler (1848-1912), later to become a renowned consulting engineer, took over the direction of the works (Schuyler 1888). In contrast to its forerunners, the upstream face of the Sweetwater dam was not vertical, but instead had an inclination similar to the down-

Abbeystead dam

First Big Bear Valley dam

Sweetwater dam

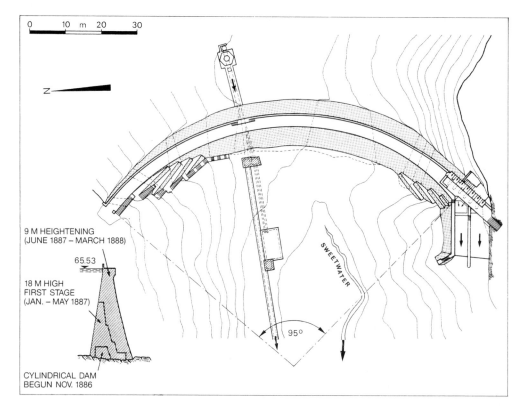

0   10 m 20   30

9 M HEIGHTENING
(JUNE 1887 – MARCH 1888)

65.53

18 M HIGH
FIRST STAGE
(JAN. – MAY 1887)

CYLINDRICAL DAM
BEGUN NOV. 1886

95°

SWEETWATER

Figure 175. Plan and cross section of the Sweetwater arch
dam completed in 1888 southeast of San Diego in California
(after Shuyler 1888).

stream side, so that in place of the upstream face radii of curvature,
those of the arch axes were constant from crest to base (Fig. 175).
A short time after its completion in 1888, the 30 m high structure
successfully withstood almost two days of overtopping by a flood,
thus testifying the inherent strength of well-built arch dams.
Nevertheless, in 1910-1911 it was transformed into a somewhat
higher gravity dam (USCOLD 1988).

6.4.2 *The breakthrough*

Pathfinder and Buffalo
Bill dams

The Sweetwater dam served as a model for, among others, the
Pathfinder and Buffalo Bill (formerly Shoshone) dams, 220 and
580 km northwest of Cheyenne in Wyoming. Built by the 'US
Bureau of Reclamation' from 1905 to 1910 these dams reached the
unprecedented heights of 65 and 99 (now 107) m respectively
(USCOLD 1988). Accordingly they required considerable quantities of construction materials, i.e. 50 000 m^3 of masonry for the
Pathfinder and 63 000 m^3 of concrete for the Buffalo Bill dam.
Even more remarkable than their dimensions, however, was the
fact that they were the first arch dams, which were not designed
merely as a stack of individual arches (Wisner & Wheeler 1905).

Their interdependence was taken into account by also considering
the median vertical section of the dam, which, statically is a cantilever fixed at the base. The water and other external loads were
then divided into vertically and horizontally carried parts by adjust-
ing the horizontal deformations of the median cantilever to the
radial deflections of the arch crowns. The latter were determined
according to the theory of elastic arches, that had finally become
generally accepted, although it still contained imperfections and
considered the arches as rigidly fixed at the abutments (Bazant
1936).

New method of arch
dam analysis

The distribution of the loads resulted in a considerable relief for
the lower arches, while the upper ones had to partially support the
cantilever. The new method of analysis had been developed in the
late 1880's by Hubert Vischer and Luther Wagoner (1847-1922)
for checking the stresses in the Big Bear Valley and the Sweetwater
dams. Unfortunately, however, it was published in a local journal
and thus escaped general attention (Vischer & Wagoner 1889).
The method became more widely known only after its publication
in 1904 by Silas H. Woodard (1870-1961), who had it reformu-
lated while designing the Cheesman curved gravity dam 62 km
southwest of Denver in Colorado, a city which, incidentally, was
also the headquarters of the 'US Bureau of Reclamation' (Harrison
& Woodard 1904).

In the discussion of Woodard's paper professor Williams (mentioned earlier), designer of the first double-curved domelike arch
dam near Ithaca in New York State, renewed the French engineer
Pelletreau's suggestion of 1879, namely to decrease the radius of
curvature of an arch dam from crest to base (Williams 1904).
Eastwood, who later became known for his multiple-arch dams
(see above), showed in a short paper in 1910, how the design of the
Buffalo Bill dam could have been improved by the new concept
(Eastwood 1910).

Variable-radius arch
dams

The decisive step was taken by Lars R. Jorgensen (1876-1938),
a native of Denmark holding German degrees in mechanical and
electrical (not civil!) engineering. In 1901 he emigrated to the USA
and worked as a hydroelectric engineer in California. He founded
his own consulting firm in San Francisco in 1914. In the same year,
construction of his first, 51 m high dam with variable radius of
curvature or about constant central angle of the arches was
completed on the Salmon Creek in southeastern Alaska (Fig. 176)
(Jorgensen 1912, 1915). Although its median vertical section re-
sembled that of the Sweetwater, Pathfinder and Buffalo Bill dams,
the shape of the Salmon Creek dam was strictly the consequence
of the new layout that became standard for variable-radius type
arch dams. The upstream bulging served to off-set the undercutting by the stronger curved lower arches near the abutments while

Salmon Creek dam

Figure 176. Plan and median vertical section of the first variable-radius arch dam completed in 1914 on the Salmon Creek in Alaska (from Creager et al. 1945).

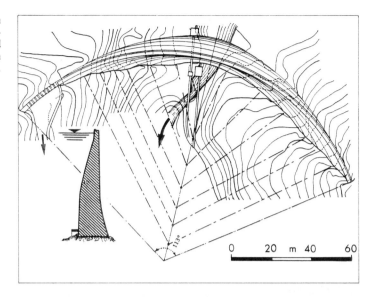

the pronounced downstream toe reduced the tensile stresses at the upstream heel. The full realization of the new design was made possible by building the dam entirely of concrete containing 280 kg/m^3 of cement. From the aggregate processing and the mixing plants near the upstream heel of the dam, the concrete was lifted to the top of a temporary steel tower and chuted into the forms.

To allow for the contraction of the concrete in its cooling off phase the Salmon Creek dam was subdivided into three construction blocks by two vertical joints – an innovation introduced at the same time into the building of gravity dams (see above). At the Spaulding dam in northern California, construction of which had begun as a curved gravity dam, whereas its upper three quarters were continued from 1913 on as a variable-radius arch dam, the contraction joints were provided with keys and filled with grout injected through drill holes in order to obtain a monolithic structure (Jorgensen 1915). In later dams special grouting pipes and outlets were embedded in or next to the joints. Moreover, their spacing was reduced to about one quarter of the 50 to 70 m used at the Salmon Creek dam.

Grouted contraction joints

The technical as well as the economic advantages of Jorgensen's design were immediately recognized. It rapidly gained popularity in the western USA and elsewhere (Table 45) (ANIDEL 1961, ICOLD 1973, 1984, Jorgensen 1915, Schnitter 1992, USCOLD 1988). Additional impetus was imparted by the accelerating pace of the development of hydroelectric power resources. Although it never replaced the cylindrical type completely, the variable-radius arch dam was to prove especially suitable for high and large structures. Scandinavian designers, the 'US Bureau of Reclamation' as

Rapid spread of variable-radius arch dams

Table 45. World record variable-radius arch dams.

Year of completion	Name of dam	Country (state)	Distance in km from city	Height (m)	Length (m)	Concrete volume (1000 m³)	Purpose
1914	Salmon Creek	USA (Alaska)	10 NW Juneau	51	195	40	Power
1919	Spaulding	USA (Calif.)	230 NE S. Francisco	84/64[1]	244	147	Power
1926	Cushman 1	USA (Wash.)	70 W Seattle	84	339	69	Power
1930	Diablo	USA (Wash.)	150 NE Seattle	119	360	268	Power
1947	Lumiei (Sauris)	Italy	115 NE Venice	136	138	100	Power
1949	Ross (Ruby)	USA (Wash.)	150 NE Seattle	165	396	695	Multi
1957	Mauvoisin	Switzerland	80 SE Lausanne	237[2]	520	2030	Power
1960	Vajont[3]	Italy	95 N Venice	262	190	351	Power
1980	Inguri	Georgia (ex UdSSR)	270 NW Tbilisi	272	680	3960	Pow./Irrig.

[1]Part above gravity base. [2]Heightened 1991 by 13 m. [3]Reservoir filled 1963 by huge rockslide; dam undamaged.

Figure 177. Downstream view of the first variable-radius arch dam in Europe, completed in 1920 near Montsalvens in southwestern Switzerland (photo by the author).

well as French engineers persisted in the use of the cylindrical design for a long time, even for large structures.

The Jorgensen type of arch dam was introduced in Europe a few years after its invention by Heinrich E. Gruner (1873-1947) (Schnitter 1992). He had inherited a renowned consulting firm in Basle/Switzerland and, ever since his stay there in 1900/1901, he closely followed the developments in the USA. In 1917 he was commissioned to design the Montsalvens power dam 43 km southwest of Berne in Switzerland. Completed in 1920, the structure was 55 m high, strongly curved and relatively sturdy (Fig. 177). Its horizontal sections were not circular but instead they followed the funicular curve for the part of the water load supported by them. Moreover, to take care of the higher stresses near the abutments the thickness of the arches increased from the crown towards them.

Montsalvens variable-radius arch dam

Both innovations resulted from the use of a considerably improved method of analysis published by Hugo F.L. Ritter (1883-1956) in 1913 (Ritter 1913). He was a son of Wilhelm Ritter (1847-1906), the well-known professor for statics at the Institutes of Technology in Riga/Latvia and Zürich/Switzerland. In the years 1909/1910 Hugo Ritter had undertaken a study tour through the USA and was familiar with Woodard's work. His method was refined by Gruner's collaborators Henri Gicot (1897-1982) and the already mentioned Stucky, who later became the two most prominent Swiss arch dam designers (Stucky 1922). In the improved analysis, the radial deflections of the arches were adjusted to those of the median vertical section as well as some additional cantilevers on both sides. This permitted the determination of the variations in the load distribution, both in the vertical and the horizontal sense.

Improved method of analysis

Ritter's method of analysis was extended at the 'US Bureau of Reclamation' under the supervision of the abovementioned Savage. The principal additions concerned the adjustment of the tangential deflections and the rotation in horizontal planes, as well as the taking into account of deformations in the foundation rock. Instead of solving the many linear equations tying the load distribution on arches and cantilevers to the adjustment of their deformations, it was, at that time, quicker to seek the proper load distribution by trial and error. This procedure, supplemented by a large body of auxiliary tables, was therefore aptly named 'Trial Load Method' (USBR 1938b). After adaptation to electronic data processing and many additional refinements, the method remained the most frequently used. However, it still represented only an approximation of the theory of thick shells, which arch dams actually are. The many attempts since the early 1920's to directly use that theory stumbled, more often than not, over the practical difficulties of its application. These were overcome only recently with the introduction of the analysis by small finite elements and the availability of computers with sufficient storage capacity and speed.

Further improvements of analysis

Parallel to these theoretical developments, great efforts were made to collect pertinent data from structures under construction or already in operation. While deflection measurements under load were initiated at the above mentioned, ill-fated Bouzey gravity dam right after its completion in 1881 (Ziegler 1900-1925), the first such measurements on an arch dam appear to have been those on the 12 m high, thin Barren Jack Creek dam in southeastern Australia in 1908-1909 (Wade 1908-1909). They were carried out from scaffoldings on the downstream face. At the Salmon Creek dam, surveying methods akin to the ones adopted at the Bouzey

Measurements on actual structures

gravity dam were used for the purpose. These methods were perfected at the Montsalvens dam and later structures. Montsalvens also was the first arch dam to be equipped with permanent instruments, i.e. embedded resistance thermometres connected by cables to the measuring stations (Schnitter 1992). In the 114 m high Spitallamm curved gravity dam – completed in 1931 some 80 km southeast of Berne in Switzerland – the first plumblines were installed in vertical shafts. They allowed simple and frequent measurements of the structure's horizontal deformations at different elevations. Replacing the weight at the bottom by a float at the top, the plumblines later could be prolonged into borings in the foundation rock.

Instrumentation of dams

On the initiative of the above mentioned Noetzli, an 18 m high cylindrical arch dam was built in 1926 on the Stevenson creek 290 km southeast of San Francisco in California for the sole purpose of studying its behaviour under closely controlled loading conditions (ASCE 1928). The measurements on the Stevenson Creek experimental dam were checked by testing its 1/40 and 1/12 scale models at the universities of Princeton/New Jersey and Boulder/Colorado respectively. The first model consisted of celluloid, whereas the second was made of a concrete similar to the one used for the prototype. In both test, the water load was simulated with a thin layer of mercury in order to obtain exaggerated and, hence, measurable deformations and strains, from which the stresses could be calculated. Although two-dimensional models of gravity dams had already been tested at the beginning of the century (Wegmann 1888-1927), the use of this technique in arch dam engineering constituted an innovation that proved highly beneficial. Later, the method evolved along two distinct lines, represented by the two laboratories that were to become the most prominent in the field. One was the 'National Laboratory of Civil Engineering' set up 1946 in Lisbon/Portugal. It used small celite-plaster models loaded with mercury. By contrast, the 'Institute for Experiments on Models and Structures' (ISMES) founded in 1951 in Bergamo/Italy, preferred relatively large pumice-concrete models to which the loads were applied by hydraulic jacks (Fig. 178).

Experimental arch dam

Parallel model tests

Perfection of model testing techniques

Involuntary 'tests' of sorts were finally provided in 1925 and 1926 respectively by the 16 m high Moyie arch dam in northern Idaho and by the 19 m high Lake Lanier (or Vaughn Creek) arch dam in western North Carolina. Both lost their right abutment over more than two thirds of their height without collapsing (ENR 1926). Thus these first failures of arch dams once more confirmed the great strength reserves inherent in this type of structure.

Partial failures of two arch dams

Figure 178. Loading test at the ISMES laboratory in Bergamo/ Italy on a model of the 180 m high Emosson arch dam built from 1967 to 1974 in south-western Switzerland (Schnitter 1974).

6.4.3 *Maturity*

Much of the great effort to clarify the structural behaviour of arch dams was undertaken in connection with the design of the epochal Hoover (formerly Boulder) dam as already mentioned. It was built as a massive curved gravity structure. This somewhat paradoxical caution on the part of the 'US Bureau of Reclamation' (Raphael 1985) initiated the decline in the construction of true arch dams in the USA. This trend is exemplified by the developments in California, where the number of new arch dams as well as their proportion in relation to the other types, decreased sharply after World War II (1939-1945) (Fig. 179).

The opposite occurred in Europe, particularly in Italy, where arch dams had been almost as popular as in California (ANIDEL 1961). Until the 1920's it primarily concerned smaller cylindrical structures located in narrow gorges. These also had relatively conservative dimensions, except for the 38 m high Corfino dam completed in 1914 some 90 km northwest of Florence. It was designed, like several other arch dams built in the following decade, by Angelo Omodeo (1876-1941). It was in this renowned consulting office where the above mentioned Marcello got acquainted with the art of dam engineering. Besides his many contiguous buttress dams, he later also designed some major structures of the arch type.

The variable-radius principle was first used in Italy for the 50 m high Gurzia dam by Gaetano Ganassini (1875-1932). The structure, located some 40 km north of Turin, was completed in 1925 with an important variation: the upstream bulging of the central part of the dam was carried so far as to result in a pronounced downstream overhang of the crest. Thus, alone for its stability

Margin notes:

Decline of arch dams in the USA

Popularity of arch dams in Italy

Gurzia variable-radius arch dam

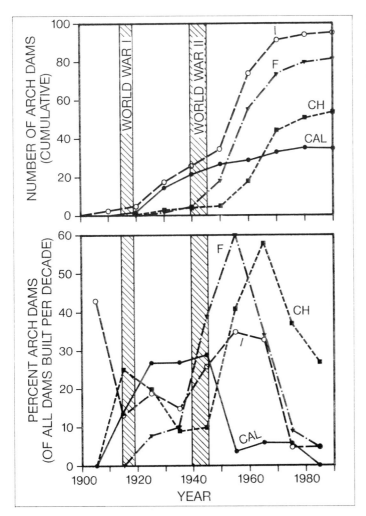

Figure 179. Development of arch dam construction between 1900 and 1990 in California (USA), France (F), Italy (I) and Switzerland (CH) (over 15 m high dams).

under the forces of gravity, it depended on arch action and the upstream heel was liable to crack and leak under the water load. This drawback was corrected by Fabio Niccolai (1892-1944) at the Osiglietta dam built 60 km west of Genoa just before World War II (1939-1945). The correction consisted of undercutting the upstream heel in order to compensate to some extent for the downstream overhang of the crest, resulting in a double-curved dome-structure similar to the much smaller one Williams had begun near Ithaca in New York State at the beginning of the century (see above) (Fig. 180).

On the basis of a model test, the Osiglietta dam furthermore was provided with a hingelike perimetric joint which separated the dome from a foundation pad, reducing the upstream tensile stresses along the abutments. This arrangement, rarely adopted outside

Osiglietta double-curved arch dam

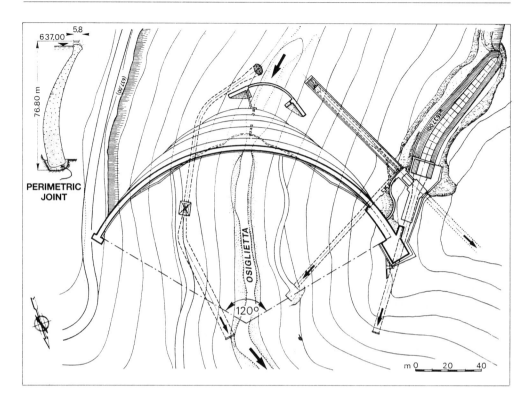

Figure 180. Plan and cross section of the Osiglietta dome dam completed in 1939 west of Genoa in Italy (from ANIDEL 1961).

Figure 181. André Coyne (1891-1960).

Italy, became characteristic of the many important arch dams designed by Carlo Semenza (1893-1961) (Semenza 1957). For a long time he was the chief engineer for hydropower plants of the 'Adriatic Society for Electricity' in Venice. His most ambitious project was the 262 m high Vajont dam built between 1956 and 1960 in a narrow gorge 95 km north of Venice. In 1963 its reservoir of 169 million m^3 capacity was to a large extent filled up by what historically is known as the largest rock-slide and which was triggered by the water impoundment. The resulting wave overtopped the dam, and downstream, killed more than 2000 people. The structure, however, suffered only little damage.

A similarly tragic end marred the career of the prominent dam designer Coyne, mentioned previously (Fig. 181). He first rose in the ranks of the French 'Corps of Bridge and Highway Engineers'. After World War II (1939-1945), Coyne founded his own consulting firm specializing in dam engineering (Coyne 1957). In his first large structure, the 90 m high Marèges arch dam completed in 1935 in the Central Massif, Coyne used the variable-radius principle with an undercut upstream heel. But, later on, especially in his high dams, he reverted to a rather sturdy cylindrical type. The latter's downstream face was defined by horizontal spirals and it was analyzed on the basis of independent arches cut from the dam by inclined planes instead of horizontal ones. The Aigle, Bort and Chastang dams, built on the Dordogne river soon after World War II (1939-1945) up- and downstream of the Marèges dam were characteristic examples for this type of design.

Simultaneously with these large structures Coyne designed many smaller arch dams, on which he experimented with new ideas, sometimes adopting extremely bold dimensions. This eventually resulted in several failures (Mary 1968). After a number of years of operation, the 47 m high Gage dam 120 km southwest of Lyon had to be replaced by a more conventional arch dam farther upstream, whereas the 90 m high, sheet-thin Tolla dam, 10 km east of Ajaccio in Corsica, had to be transformed into a curved gravity structure soon after its completion. The disastrous 1959 collapse of the 66 m high Malpasset dam, 45 km southwest of Nice cost 421 lives when the poor foundation rock gave way on the left flank upon the first complete filling of the reservoir. To this date it has remained the only total failure of an arch dam in the world. Like most such events, it taught lessons that still were to be learned. It gave a great impetus to the young science of rock mechanics. The first book on the subject had been published just two years earlier by Joseph A. Talobre (born 1903) (Talobre 1957). Moreover, arch dams, like gravity dams, henceforth were provided with drainage arrangements. However, these mainly served to protect the downstream part of their mostly steep abutments from uplift pressures, rather than the structures themselves that were hardly affected.

After the Malpasset disaster and the death of Coyne, the proportion of new arch dams built in France decreased sharply (Fig. 179). By contrast the great development which had begun in Switzerland after World War II (1939-1945), continued through the 1960's. The great majority of the Swiss arch dams were of the variable-radius type with a moderate vertical curvature. The most important structure was the 237 (now 250) m high Mauvoisin dam 80 km southeast of Lausanne, completed in 1957 to the design of Stucky (Stucky 1956). Noteworthy among the smaller structures is Gicot's 51 m high Vieux Emosson dam constructed in 1954 and 1955 some 50 km south of Lausanne, as it embodied the first use of parabolic arches to flatten their curvature near the abutments (Gicot 1961). For the 86 m high Les Toules dam completed in 1963 about 80 km southeast of Lausanne, Gicot introduced elliptical arches, whereas his 125 and 93 m Hongrin twin arch dams, finished in 1969 approximately 30 km east of Lausanne, were outstanding in respect of both originality and elegance (Fig. 182).

The 1960's also brought a revival of the arch dam construction in the USA, particularly so after the adoption of the double-curvature design by the 'US Burea of Reclamation' for some of its most important structures like, for example, the 153 m high Flaming

Figure 182. Aerial view of the Hongrin twin arch dams in southwestern Switzerland (photo EOS, Lausanne).

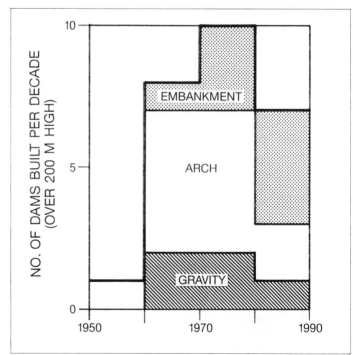

Figure 183. Very high dams of the three main types completed in the world between 1950 and 1989.

Gorge dam completed in 1964 in northeastern Utah, the 160 m high Yellowtail dam finished in 1966 in southern Montana or the 143 m high Morrow Point dam completed in 1968 in western Colorado. Furthermore, arch dams became established in such regions of great dam building activity as the Iberian peninsula, Japan and the former Soviet Union (CIS) as well as generally spread around the globe. At the same time, both the designs and the

Table 46. No. of over 100 m high arch dams in the world.

| Year of completion | Maximum height of dams (m) | | | | | Industrialized countries[1] |
	100-149	150-199	200-249	250-300	100-300	
1920-1929	2				2	2
1930-1939						
1940-1949	5	1			6	5
1950-1959	21	6	1		28	24
1960-1969	33	13	4	1	51	45
1970-1979	17	6	5		28	21
1980-1989	10	5	1	1	17	9
Totals	88	31	11	2	132	106

[1]Industrialized countries: Australia, Canada, CIS, Europe, Japan, New Zealand and USA.

construction techniques reached considerable uniformity. But in spite of the high perfection and productivity attained in their construction, arch dams lost ground in the last decades. Similarly to the classical gravity dams they were outranked by embankments, even in the field of very high structures (Fig. 183, Table 46).

Arch dams world-wide

6.5 FOUNDATION TREATMENT

Although the abovementioned first director of the school for the French 'Corps of Bridge and Highway Engineers', Perronet, had recommended the investigation of foundation conditions by test borings and pits ahead of construction as early as 1769, it took more than one century before his advice was followed. Obviously, there still were plenty of favourable dam sites, mostly on competent rock, which could satisfactorily be judged by visual inspection and on the grounds of experience. Moreover, most dams were still of moderate height. One of the first instances where both conditions were no longer met, was the abovementioned Vyrnwy gravity dam. It had to be founded up to 19 m (almost half of its height, see Fig. 151) below the valley floor which was carefully investigated before construction started in 1882 by means of 13 pits and 177 borings (Ziegler 1900-1925). The latters' type is not reported, but basically two methods still in use were already available at the time, namely the originally Chinese wash-boring with casing, chisel and a sampling tool in loose soils and, in rock, the diamond drill with core recovery invented by the Swiss watch-maker George Auguste Leschot (1800-1884) in 1863 (Marx 1990). Nowadays, very often the second method is also used in loose soils. The interpretation of the results of a foundation investigation, which later mostly included galleries driven into the abutments, became a common task of the civil engineer and the geologist as well as the

Foundation investigations

Boring techniques

geotechnician, who tested the samples and cores taken to his laboratory.

It was already clear to the Roman engineers that a dam needed to be tied into the foundation by excavating it to a certain depth (see e.g. Fig. 62). The removal of the overburden under the upstream heel of the earth dams in the German Harz Mountains has already been mentioned earlier. It may well have been there, at the Oder dam built from 1715 to 1722, that gunpowder placed in plugged chisel-drill-holes was first used for excavation work and for the breaking-up of boulders too large for the dry masonry shells (Schmidt 1989). Removal of the material was done by shovel and wheelbarrow. Essentially, the same methods were still employed from 1861 to 1866 at the abovementioned Gouffre d'Enfer gravity dam in France. In deference to its then extraordinary height of 60 m, the foundation was prepared very carefully (Graeff 1866). After removal of all weathered rock, the foundation surface was roughened by superficial blasting and all fissures in the rock were sealed with mortar (Fig. 184).

In the following years the efficiency of excavation works was greatly increased by the invention of dynamite in 1863 by the Swedish engineer Alfred Nobel (1833-1896) and the perfection of the percussion drill driven by compressed air two years later in the USA (Sandström 1963). At the same time, steam powered shovels found general use in the USA. Without such equipment and a

Foundation preparation

Figure 184. Construction view of the Pérolles gravity dam in Switzerland in 1871, with very little machinery as yet to be seen (photo EEF, Fribourg/FR).

network of temporary narrow-gage railroads, the enormous excavation volume of 1.4 million m³ of river deposits and 0.3 million m³ of rock for the New Croton gravity dam north of New York could hardly have been accomplished in less than 4 years (Table 37) (Wegmann 1888-1927). It reached 37 m or 40% of the dam's height below the river bed and its total volume exceeded three times that of the dam!

Foundation excavation

One way of avoiding such huge excavations at dam sites with considerable overburden consisted in the lowering of steel and, later, concrete caissons, in which an air pressure equal to the exterior water pressure was maintained during excavation. Originally developed in the 1830's for the sinking of mine shafts and later, of bridge piers, the method was adopted by the Swiss expert in this domain, Conradin Zschokke (1842-1918) (Vischer & Schnitter 1991). He applied it in 1897 for the foundation of the Hagneck gated weir 23 km northwest of Berne in Switzerland. Later, many more such weirs were founded by this method in Switzerland and abroad until it became obsolete in the 1950's (Fig. 185).

Caisson foundation

Figure 185. Caisson foundation in 1912 for the Laufenburg barrage-dam on the River Rhine, 42 km northwest of Zürich; above before and below after lowering of the caisson (photo KWL, Laufenburg/AG).

Foundation grouting

Still another method of avoiding deep excavations was the improvement of the subsoil by means of injections of cement or chemicals through boreholes (Glossop 1960-1961). Invented in 1802 by the French engineer Charles Bérigny (1772-1842) during harbour works, the method was first used in 1879 for dam engineering by Hawsley (mentioned above) to seal fissures in the rock beneath the 25 m high, leaky Tunstall dam, approximately 30 km southwest of Newcastle u/T (Binnie 1987). He again applied the method in 1879-1880 and 1885-1886 to stop leakage beneath the 21 m high Cowm embankment, 22 km northeast of Manchester. At the abovementioned New Croton dam, grouting was used to fill cavities in the limestone (Wegmann 1888-1927).

Grouted cut-offs in the USA

A grouted cut-off was first incorporated in the design of the 25 m high River Mill (formerly Estacada) flat-slab buttress dam, built in 1910-1911 on highly permeable volcanic breccia 40 km southeast of Portland in Oregon (Glossop 1960-1961, USCOLD 1988). The curtain along the dam's heel consisted of two parallel rows of diamond-drilled holes spaced 1.8 m apart and of 15 m average depth. They were tested by water pressure prior to the injection of grout by means of a machine operated by compressed air. The design was immediately adopted by the 'US Bureau of Reclamation' for its 47 m high Lahontan earth dam, 70 km northeast of Carson City in Nevada, and again at the 107 m high Arrowrock gravity dam, as already mentioned (USCOLD 1988). Up to 1930, rock grouting was used on 19 dams in the USA (Glossop 1960-1961).

European practice

In Europe, a two-row grout curtain was installed for the first time in 1921-1922 at the 79 m high Barberine gravity dam in Switzerland, 50 km south of Lausanne (Fig. 186) (Hugentobler 1927). Later, one-row curtains (sometimes with two intersecting systems of drillholes) were deemed sufficient in Europe, but much

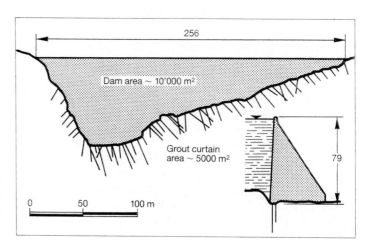

Figure 186. Longitudinal and cross sections of the Barberine dam in Switzerland showing the grout holes executed in 1921-1922 (after Hugentobler 1927).

Figure 187. Upstream view of a model of the grout curtain extending upstream from the 234 m high El Cajon arch dam in northwestern Honduras, completed in 1985; its construction required 11.5 km of tunnels and 535 km of drill-holes (photo: Rodio, Melegnano/I).

higher grouting pressures were used than in the USA. Wherever these curtains needed to be extended upstream from the dam, they reached considerable dimensions (Fig. 187) (Schnitter 1984). The world record is presently held by the $1\ 200\ 000\ m^2$ curtain under 184 m high Atatürk rockfill dam, built from 1983 to 1990, on the Euphrates river in southeastern Turkey. Its construction required over 10 km of tunnels and 1200 km of drill-holes (Ilker & Özbay-oglu 1991).

Large grout curtains in rock

In the meantime, the problem of establishing grout curtains in alluvial soils was tackled by an invention of the Swiss engineer Ernest Ischy (1905-1977) in connection with the 55 m high Bou-Hanifia bitumen-faced rockfill dam constructed from 1933 to 1940 about 320 km southwest of Algiers (Glossop 1960-1961). The invention consisted in the introduction into the drill-hole of a tube with holes every 0.3 m. These were covered on the outside by tightly fitting rubber sleeves. By means of a second, smaller tube provided with a double packer at its end, the grout could subsequently be injected individually through the holes and sleeves along the outer tube at the desired depths. After washing, these tubes and sleeves could be reused several times for additional grouting. At Bou-Hanifia, this technique was employed only in the substantial lateral extensions of the concrete cut-off wall beneath the upstream heel of the dam. The first grout curtain in alluvium beneath an embankment was built from 1955 to 1957 for the 129 m high Serre-Ponçon central core earth dam, 150 km northeast of Marseilles in France. It had several parallel rows of drill-holes and a depth of up to 115 m. Under the 111 m tall Aswan High Dam, completed 1970 across the River Nile 700 km south of Cairo, an alluvial grout curtain even reached a depth of 208 m and sealed an area of $57\ 000\ m^2$ (Blatter & Lendi 1968).

Grout curtains in alluvial soils

Important alluvial grout curtains

Development of slurry
walls

In the last decades, alluvial grouting was replaced by 0.6 to 1.0 m thick walls of interlocking piles, and later on, panels excavated in a bentonite slurry, that supported the soil and which, after completion of the excavation, was displaced by the raising concrete, clay and earth or bituminous material of the wall. The method, developed by the Austrian engineer Christian Veder (1907-1984), was applied for the first time in 1951 on a 28 m deep cut-off, under the heel of the Colle Torcino gated weir, 140 km southeast of Rome (D'Andrea 1952). Thereafter it spread rapidly in Italy and abroad and found its way also into general foundation work. The depth record of 123 m was attained 1990 by a slurry wall built into the leaky core of the 42 years old Mud Mountain rockfill dam in Washington State (Table 35). From 1984 to 1992, up to 45 m deep

Record walls

walls with an extraordinary total area of 920 000 m² were constructed underneath two thirds of the 70 km long (3 km more than the 17th century Hongze dam in China) Yacyreta dam on the Argentine-Paraguayan border, 250 km south of Asunción (Garbe et al. 1991).

6.6 SPILLWAYS

6.6.1 *Hydrological bases*

Discharge measurements

For the proper design of a spillway the flood discharge of a river to be dammed needs to be established. Although river stages were observed on the Nile since about 2200 BC, it was not until the universal genius Leonardo da Vinci (1452-1519) for the first time determined the velocity of the flow in a river with a float and, by multiplying it with the wetted cross-section, correctly computed its discharge (Frazier 1974). In the 17th century, the floats were replaced by paddles and ultimately by paddle wheels, the precursors of the modern current metres. At the same time, the concept of the hydrologic cycle of evaporation, cloud formation, precipitation and run-off, finally became generally accepted and was applied also quantitatively (Biswas 1970). The first systematic computations of daily discharges were carried out, on the basis of the stages recorded from 1809-1820 in the River Rhine at Basle/ Switzerland, by Hans Conrad Escher (1767-1823), former manager of the Linth reclamation project in northeastern Switzerland (1807-1816) (Vischer 1983).

Flood estimation

For as long as only scant discharge data were available, flood flows were estimated by so-called 'rational' formulas on the basis of maximum daily rainfall, a run-off coefficient and the catchment area (Biswas 1970). The originator of the method was the Irishman Thomas J. Mulvaney (1822-1892). As more flood discharge data became available numerous formulae were developed, by means of which the flood discharge in a given region or even on a world-wide basis could be determined in function of the square or some other root of the catchment area and a corrective coefficient (Biswas 1970). The first such formula was developed by colonel C.H. Dickens 1865 on the basis of his observations in India. The increasing amount of data led the American engineer George W. Rafter (1851-1907) to suggest in 1896 their statistical evaluation according to the normal law of errors of Karl F. Gauss (1777-1855) (Fuller 1914). In 1914, Allen Hazen (1869-1930) discovered in the USA that the logarithms of the discharges fitted this law better than their numbers. Other methods of frequency analysis were later proposed by many hydrologists.

    Meanwhile the American Le Roy K. Sherman (1870-1954) demonstrated the use of the unit hydrograph for translating rainfall excess into a flood hydrograph in 1932 (Sherman 1932). In the 1940's the 'US Army Corps of Engineers' and the 'US Weather Bureau' jointly started to determine the magnitude and temporal as well as spatial distribution of the probable maximum precipitation for given areas on the basis of worst possible meteorologic conditions, from which the probable maximum flood could be determined.

Flood formulae

Flood frequency analysis

Probable maximum flood

6.6.2 *Overflows*

The simplest way to dispose of excess flood waters was to let them flow over the crest of the dam, provided it was built of masonry or concrete and not too high. At embankment dams the danger of rapid erosion by the overflowing water usually precluded such a solution. At high masonry dams, there was the danger that scouring of the downstream river bed by the impact of the falling water could undermine the dam's toe and/or the valley sides. Therefore, already in some ancient (Figs. 20, 44) as well as in most Moslem (Figs. 75 to 79) dams, the downstream faces were inclined in order to guide the overflowing water back into the river, a provision that, incidentally, yielded also a statically advantageous dam profile. Since their inclination was constant, these faces however were subject to the formation of subatmospheric pressure under the overflowing water and to ensuing cavitation damage in the masonry. The Italian hydraulician Giorgio Bidone (1781-1839) first attempted to formulate the correct parabolic shape, while the member of the French 'Corps of Bridge and Highway Engineers', Henri E. Bazin (1829-1917), studied it 1886/1888 by means of very extensive laboratory tests (Rouse & Ince 1957). One of the first large dams built with a parabolically shaped overflow was Vyrnwy (Fig. 151).

Flood discharge over dam crest

Correct shape of overflow

Separate overflows

Downstream chutes

Figure 188. Cross sections of one of the two side channel spillways at Hoover dam (above, after USBR 1938a) and of the I.H. Juanda (Jatiluhur) morning glory spillway (below, after Bohn & Hamon 1967).

Overflows at one or both ends of a dam, or over depressions along the reservoir rim, were used at many ancient (Figs. 27, 37, 53) and medieval (Fig. 90) embankment dams. From the above mentioned, postmedieval power dams in the German Harz Mountains onwards, channels or chutes were often provided downstream of the overflows to guide the water back into the river (Schmidt 1989). The crest of the overflow was sometimes curved in plan to increase its discharge capacity, as was the case at the Rudyard navigation dam completed 1797 in Great Britain (Fig. 125).

Another method to accommodate a long overflow was to build it along the reservoir rim and to collect its discharge in a side channel carrying it downstream around the end of the dam, quite often in a tunnel. Some of the earliest installations of this kind were used 1866-1867 at the Gouffre d'Enfer and Ternay flood retention dams, 6 and 25 km respectively southeast of St. Etienne in eastern France (Ziegler 1900-1925). Their discharge capacity still was relatively modest, but at the New Croton dam, built from 1892 to 1906 north of New York, a side channel spillway, able to discharge 3000 m^3/s, was provided (Wegmann 1888-1927). The largest of

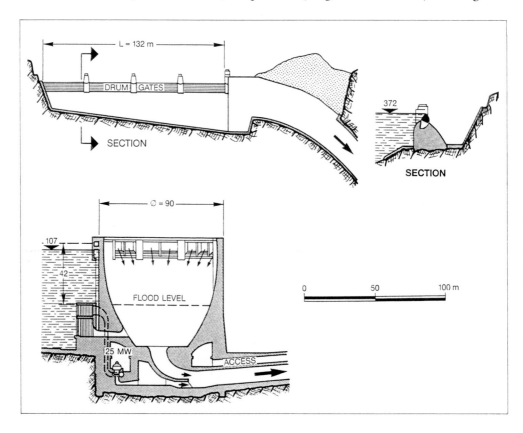

their kind so far constructed are those on either side of the Hoover (formerly Boulder) dam. They are capable of discharging up to 5700 m^3/s each and they had been extensively tested on hydraulic models prior to their construction (Fig. 188, above) (USBR 1938a).

A still further way of lengthening an overflow was to place it on top of a vertical, circular shaft. Such shafts had been used as outlets since Antiquity (Figs. 28, 38, 57, 59) and the Middle Ages (Figs. 82, 83), but the first use of a so called 'morning-glory' spillway appears to have occurred in 1887 at the 16 m high Torcy Neuf navigation dam, 120 km north of Lyon in eastern France (Ziegler 1900-1925). It had a diameter of 3.2 m and, hence, a discharge capacity of just about 40 m^3/s. 120 m^3/s were already attained in 1908 at the Rodannenberg (or Klöntal) embankment dam, 50 km southeast of Zürich in Switzerland (Ehrensperger 1910). In 1929, the Gibson arch dam, 120 km north of Helena in Montana, was equipped with a shaft spillway discharging over 1400 m^3/s (USCOLD 1988). The largest of them all is the one at the I.H. Juanda (Jatiluhur) dam, completed in 1967 some 80 km southeast of Jakarta in Indonesia (Fig. 188, below) (Bohn & Hamon 1967). With a diametre of 90 m it can discharge 3000 m^3/s and its base contains even six 25 MW turbines!

Side channel spillways

Shaft spillways

Record structures

6.6.3 *Overflow control*

The installation of gates to control the overflow became mandatory whenever large quantities of water had to be spilled without raising the storage level too much during floods. Simple wooden slide gates were at first used for the purpose, but in the 1770's, shutter or flap gates, which were hinged to the top of the dam and held by removable props on the downstream side, were developed in southern France for the navigation on the river Orb, 180 km west of Marseilles (Wegmann 1888-1927). A century later, chains were introduced to operate the flaps from a bridge. It was to remain the preferred method for a long time, although the famous French turbine designer L. Dominique Girard (1815-1871) had already proposed the use of hydraulic jacks in 1869, as they are customary today (Fig. 189) (Wegmann 1888-1927). At present, the largest flap gate in the world is installed, since 1965, at the St. Pantaleon weir 135 km west of Vienna in Austria (Erbiste 1981). It is 100 m long, 3.7 m high and thus supports a hydrostatic load of some 7 MN.

The Irish engineer Francis G.M. Stoney (1837-1897) developed a plane steel gate around 1875. It moved on trains of rollers in vertical grooves and rapidly became very popular (Wegmann 1888-1927). The first large gates of this type were installed in 1883 for the Lough Erne drainage works, 140 km west of Belfast in

Need for gates

Flap gates

Operation by hydraulic jacks

Roller gates

Figure 189. Flap gate operated by a hydraulic jack as designed in 1869 by L.D. Girard in France (after Wegmann 1888-1927).

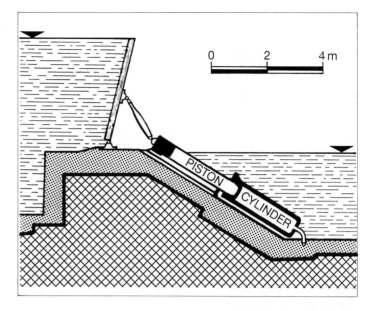

Ireland, and supported a hydrostatic load of 0.9 MN each. 22 MN were attained in 1914 at the Laufenburg weir on the High Rhine, 40 km northwest of Zürich/Switzerland (Schnitter 1992). Its up to 16.0 m high and 17.3 m long gates were subdivided into two leaves (Fig. 190). These could be lowered (overflow) or lifted (underflow) individually, but they also could be removed altogether from the water so as to let the maximum discharge pass. Today, the largest wheeled gates resisting 25 MN each, are the eight installed in 1979 at the La Grande 2 dam, 900 km northwest of Quebec in Canada (Erbiste 1981).

Evolution of roller gates

In the last decades, however, the plane gates have given way to the radial ones, the damming surface of which is a segment of a cylinder rotating about a downstream horizontal axis. One of the earliest uses of this type of gate occurred in 1853 on the Seine river in central France by C.A. François Poirée (1785-1873), an inspector of the French 'Corps of Bridge and Highway Engineers' (Rhone 1959). In the 1860's, a former chief engineer of the corps D. Eugenie Mougel (1808-?) designed similar steel gates with upstream axes for the left half of the Delta barrage, 16 km northwest of Cairo in Egypt. They withstood 0.6 MN of hydrostatic load each. The loads exceeded 40 MN shortly after World War II (1939-1945), when at times, flaps were added on top of the radial gates for the passage of floatsome and ice (Erbiste 1981). Moreover, hydraulic jacks became customary for the operation of the flaps as well as for the gates themselves (Fig. 191).

Radial gates

Development of radial gates

The jacks and the traditional cable or chain hoists depended on a power supply, the provision of which could not always be

Figure 190. Downstream view of one leaf of the Laufenburg wheeled gates during erection; the upstream skin of the gate is still missing (photo KWL, Laufenburg/AG).

guaranteed. Therefore, automatic gate types were developed. They were lowered by the excess pressure of the overflowing water and could again be lifted by a backpressure equal to the storage water level after passage of the flood. The earliest such device was the 'bear-trap' gate invented in 1818 by one Josiah White from Philadelphia in Pennsylvania (Wegmann 1888-1927). It consisted of two flaps leaning against each other, the space between them being filled with water connected to the reservoir (Fig. 192, top). As soon as the water level in the latter rose the upstream flap folded down over the downstream one. Bear-trap gates remained relatively small and reached a largest hydrostatic load of 5 MN (Erbiste 1981).

Of late, the bear-trap gate was in a certain way displaced by the inflatable weir, the first prototype of which was built in 1957 by Norman M. Imbertson on the Los Angeles river in Los Angeles/

Automatic gates

'Bear-trap' gates

Inflatable weirs

Figure 191. Three of the nine radial gates in the spillway of R. Leoni (Guri) gravity dam in Venezuela; each gate carries a hydrostatic load of 33 MN and can discharge 4450 m³/s (photo W. Nüssli, Oberrohrdorf/AG).

California (Imbertson 1960). The device consists of a tube of rubberized fabric, inflatable by air or water pressure to up to 5 m of height and fixed along its base. For the passage of floods it can easily be deflated.

Hiram M. Chittenden (1858-1917) of the 'US Army Corps of Engineers' in 1896 invented the drum or sector gate, consisting of a floating body hinged up- (drum) or downstream (sector) to the top of the dam and being lowered into an appropriate chamber by the overflowing water (Fig. 192, bottom) (Wegmann 1888-1927). The chamber is connected to the reservoir as in the bear-trap gate, but in contrast, the drum or sector gate attained considerable sizes soon after its invention. Today, the largest supports a hydrostatic load of some 33 MN (Erbiste 1981).

A fourth device used for the automatic spilling of excess water was the siphon. The first two of its kind were installed in 1867 by

Drum and sector gates

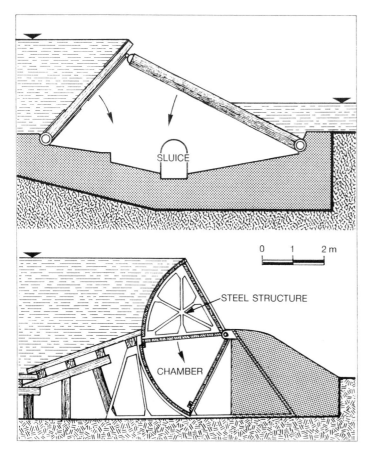

Figure 192. Cross sections of an early French bear-trap gate (top) and a Chittenden sector gate (bottom, both after Wegmann 1888-1927).

the French engineer Joseph Hirsch (1836-1901) at the Mittersheim reservoir, 65 km northwest of Strasbourg (Rothmund 1962). They consisted of steel pipes and discharged 6.5 m³/s each. At the turn of the century, the Italian engineer E. Gregotti started to build siphons of reinforced concrete, e.g. 1901 at the Lago Lungo dam, 20 km northwest of Genoa (Fig 193, left) (ANIDEL 1961). Their discharge capacity was increased rapidly and reached 57 m³/s for each of the six siphons that were added in 1916 to the Sweetwater arch dam (see above) (USCOLD 1988). In 1946, a unique peak was attained at the Hirebhasgar dam, 320 km northwest of Bangalore in Karnataka/India, where eleven combinations of siphon and shaft spillway invented by one V. Ganesh-Iyer were built. They discharged 260 m³/s each and some 2900 m³/s together (Fig. 193, right) (Doddiah 1955). More recently, siphon spillways have become rather unpopular because they discharge immediately at full capacity and thus might cause otherwise avoidable flooding downstream.

A fifth automatic spilling device, sometimes provided to release

Siphon spillways

Development of siphons

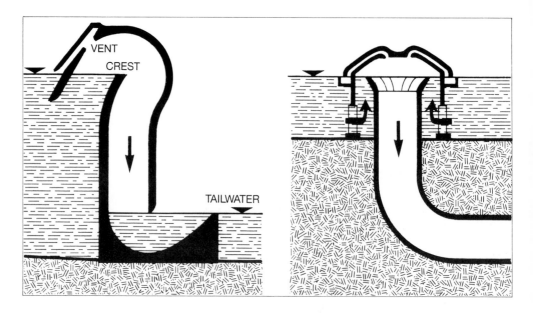

Figure 193. Cross sections of a Gregotti siphon spillway (left, after ANIDEL 1961) and of the combination of siphon and shaft spillway invented by V. Ganesh-Iyer (right, after Doddiah 1955).

discharges in excess of the design flood, is the so called 'fuse plug' i.e. an embankment designed to fail by overtopping and erosion prior to the main dam doing so. The same purpose may be achieved by 'fuse gates' arrested by bolts designed to break at a given surcharge or designed to be overturned by it.

6.6.4 *Energy dissipation*

Need for energy dissipation

Since dam heights and spillway discharges increased rapidly as from the last quarter of the 19th century onwards, means had to be found to dissipate the energy of the overflowing water that would otherwise endanger the dam's toe and/or the valley sides by erosion (Table 47) (Brown 1983-1984, ICOLD 1973, 1984, USCOLD 1988). An old but rather rarely used method, was to create a water cushion sufficiently deep to absorb the impact of the overflowing water at the toe of the dam. Thus, the older parts of the already mentioned Pontalto arch dam in South Tyrol were protected by the reservoir of the similar Madruzza dam, which was built downstream in 1883. On the basis of some sorry experiences, the effectiveness of such water cushions was recently questioned (Häusler 1980). The classical example of what could happen, was the 400 000 m^3 scour hole eroded within a few years downstream of Kariba arch dam, 130 km south of Lusaka in Zambia (Fig. 194).

Downstream water cushion

Energy dissipation methods were more effective which involved a considerable entrainement of air by the overflowing water. One of these was the stepped or cascade chute as used e.g. 1866 below the outlet of the flood diversion canal around the

Table 47. World record spillways (over 1000 m³/s discharge capacity).

Year of completion	Name of dam	Country (state)	Distance in km from city	Dam type	Spillway discharge (m³/s)	Energy dissipated (1000 MW)
1871	Ekruk	India	350 SE Bombay	Embankment	*1240*	*~0.2*
1871	Madhopur	India	440 NW Delhi	Boulder weir	*6230*	*~0.1*
1874	Barden (Low.)	Great Britain	60 N Manchester	Embankment	1350	*~0.3*
1878	Tajawala	India	190 N Delhi	Boulder weir	*9340*	*~0.2*
1879	Khadakvasla	India	110 SE Bombay	Gravity[1]	2500	*~0.9*
1885	Paricha	India	370 N Delhi	Gravity	*23000*	*~2*
1907	Great Falls	USA (S.C.)	60 N Columbia	Gravity	16600	*~4*
1913	Medina	USA (Tex.)	140 SW Austin	Gravity	18600	*~7*
1928	Conowingo	USA (Md.)	60 NE Baltimore	Gravity	*33100*	*~4*
1931	Sukkur	Pakistan	360 NE Karachi	Gravity	*42300*	*~1*
1936	Hoover (Boulder)	USA (Ariz./Nev.)	42 SE Las Vegas	Gravity	11300	*19*
1937	Bonneville	USA (Oreg./Wash.)	50 E Portland	Gravity	*45300*	*4*
1942	Grand Coulee	USA (Wash.)	255 E Seattle	Gravity	28300	*27*
1943	Shuifeng	China/Korea	720 E Beijing	Gravity	37500	*~35*
1955	Samara	Russia	800 SE Moscow	Gravity	*55000*	*8*
1957	The Dalles	USA (Oreg./Wash.)	110 E Portland	Gravity	*64800*	*10*
1967	Mangla	Pakistan	190 N Lahore	Embankment	56100	*46*
1970	Farakka	India	250 N Calcutta	Gravity	*70500*	*~1*
1982	Itaipu	Brazil/Paraguay	310 E Asunción	Buttress	61400	*53[2]*
1984	Lhano (Tucurui)	Brazil	1400 N Brasilia	Gravity	*110000*	*53*

[1]Downstream earth backing added after impounding. [2]Is over four times capacity of power plant.

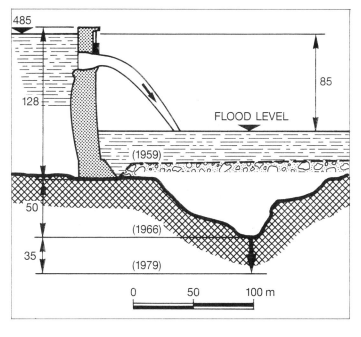

Figure 194. Cross section of the Kariba arch dam in Zambia/Zimbabwe and its downstream scour hole, which in 1979 reached a depth of 85 m but fortunately only a limited extension across the valley (after Häusler 1980).

Cascade chute

Use of hydraulic jump

Stilling basin

Flip bucket

reservoir of the abovementioned Gouffre d'Enfer dam in eastern France (Ziegler 1900-1925). The downstream face of the overflow into the described side channel spillway at the New Croton dam north of New York also was stepped (Wegmann 1888-1927). The technique attracted renewed interest in connection with the recent renaissance of gravity dams built of roller compacted concrete (see above).

Another method of energy dissipation was that of the creation of a standing wave or hydraulic jump, which forms at the transition from rapidly flowing water (as in a chute) to a slower river flow. The first to study this phenomenon, was the abovementioned Bidone (Rouse & Ince 1957). Also, the French engineers Darcy and Bazin (already mentioned) carried out extensive experiments between 1855 and 1865, on the basis of which Bresse was able to formulate the correct equation for the hydraulic jump in 1868. He ascribed it to his predecessor as professor of mathematics at the school of the French 'Corps of Bridge and Highway Engineers', Belanger (Hager 1990). To protect the river bed against erosion at the location of the jump and in order to contain the latter, so called 'stilling basins' were built, mostly of reinforced concrete. By adding sills, baffles and other obstructions the basin length could be shortened, but the additions quite often were damaged by abrasion and/or cavitation (Fig. 195).

Obviously, the most effective air entrainement was achieved by throwing the overflowing water into the air and as far downstream as possible by means of a 'flip bucket' at the lower end of the spillway chute. One of the earliest of these devices was provided in 1890 at the Vyrnwy dam repeatedly referred to above (Fig. 151). A most spectacular use of them was made by the abovementioned

Figure 195. Bird's-eye view of the model of a stilling basin with baffles after completion of scouring tests (photo Lab. Hydr. Research, Swiss Fed. Inst. of Technology, Zürich).

Figure 196. Ski jump-spillways on the roof of the Aigle power-house in France in operation (photo EdF, Paris/F)

French dam designer Coyne soon after World War II (1939-1945) at his dams on the Dordogne river in central France. He aptly named them 'ski jump' spillways and sometimes even arranged them on the roof of the power house at the dam's toe (Fig. 196).

6.7 OUTLETS

To divert the required amount of water at any time controlled outlets were necessary in most dams. From the postmedieval Spanish irrigation dams onwards, separate bottom outlets were often provided, either to purge sediment deposits from the reservoir, for the additional passage of flood waters, or just to lower the storage water level in case of need (structural defects, times of war etc.) The Spanish bottom outlets had large cross sections that were closed by boards; these were dangerous to remove and could only be put back in place after the complete emptying of the reservoir (Figs. 110, 122). In contrast later dams usually had smaller outlets equipped with gates or valves to control their discharge.

Originally, these outlets consisted of galleries (Figs. 121, 124), then of steel pipes installed in galleries and, since the end of the 19th century, of steel pipes embedded in the dam's masonry (Ziegler 1900-1925). At embankment dams, and frequently, for the diversion of the river during construction, tunnels were built around the dam site that could subsequently be used as bottom outlets. Their discharge today is almost exclusively controlled by slide, wheeled or radial gates similar to those adopted in spillways. By comparison a great variety of valves was formerly employed and this is still the case on the outlets for the useable water required. Some types of

Two kinds of outlets

Earliest bottom outlets

Outlet tunnels

Gates and valves

Energy dissipation
below outlets

valves also dissipated the energy of the outflow by air entrainement, like those of the hollow-jet, needle or fixed-cone type (Fig. 197).

By contrast stilling basins or flip buckets were required downstream of many of the bottom outlets, which, like the spillways, attained considerable dimensions in the 20th century (Table 48) (Erbiste 1981, ICOLD 1973, 1984, INCOLD 1979, Schnitter 1992). Sometimes, rapid extrapolations lead to trouble. This was for example the case in 1959 at the 226 m high Bhakra gravity dam, then under construction 240 km north of Delhi in India (Malhotra 1960). Intensive vibrations of the temporary wheeled gates installed in a diversion tunnel around the dam site caused the floor of the operating chamber to fail, killing 10 workers and flooding the

Outlet failure

Figure 197. Three modern fixed-cone outlet valves in action (photo Noell GmbH, Würzburg/D).

Table 48. World record bottom outlets.

Year of comple-tion	Name of dam	Country (state)	Distance in km from city	Hydr. load on one gate (MN)	Total discharge (m³/s)	Energy dissipated (1000 MW)
1902	Aswan	Egypt	700 S Cairo	2[1]	*14 500*	*~1.5*
1910	Nisellas	Switzerland	105 SE Zürich	7	600	~0
1948	Génissiat[2]	France	80 E Lyon	*62*	2 200	1.4
1957	Hirakud	India	470 W Calcutta	7	*29 800*	*8.9*
1958	Picote	Portugal/Spain	190 E Porto	*66*	2 100	1.9
1963	Mangla	Pakistan	190 N Lahore	*69*	11 400	*9.7*
1967	Saratov	Russia	730 SE Moscow	23	*58 400*	2.5
1975	Leoni (Guri)[2]	Venezuela	510 SE Caracas	67	17 000	*14.6*
1976	Tarbela	Pakistan	320 NW Lahore	*75*	6 200	8.6
1982	Itaipu[2]	Brazil/Paraguay	310 E Asunción	*186*	35 000	*16.5*

[1]After second heightening 1934: 4 MN. [2]River diversion outlets; gates used only once for closure.
Italics: Record breaking dimensions.

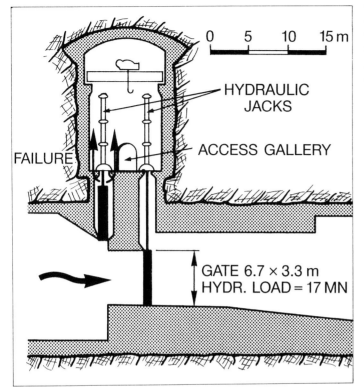

Figure 198. Longitudinal section of the temporary gate chamber in a diversion tunnel around Bhakra dam in India; the gates are shown in their normal position; at the time of the failure, the upstream gate was almost completely lowered (after Malhotra 1960).

power house after five 90 MW-generators had already been installed (Fig. 198). In order to choke the uncontrolled flow, large quantities of concrete boxes, sand and clay were dumped into the 84 m deep water in front of the upstream portal of the tunnel. The latter was then plugged as originally foreseen.

Generally speaking, gate vibrations in bottom outlets as well as in spillways could be prevented by carefully designing the lower lip of the gates and by an ample supply, through separate conduits, of air to their downstream side. Since bottom outlets controlled the whole – sometimes very important – reservoir content, they were mostly provided with two gates in series (Fig. 198). Whereas the downstream one served the normal operation, the upstream 'emergency' gate was lowered only in the event where the 'operating' gate suffered damage or was obstructed by floatsome, sediments etc. Quite often, the two gates were of different design.

Details of outlet gates

CHAPTER 7

Conclusion

7.1 STATE OF THE ART

Based on the evolution described in the preceding chapter and due to the great construction activity since World War II (1939-1945), dam engineering has become a mature technology (Fig. 199). As illustrated by the recent development of the roller compacted concrete gravity dam this fact does not preclude the potential for further genuine innovations. Quite to the contrary, considerable progress is still expected in the fields of computerized analysis, materials technology, construction techniques and monitoring of dam behaviour (see below).

Mature, but not stagnant technology

The ever increasing number of publications, textbooks and specialized periodicals in various languages, as well as the regular meetings and congresses of an 'International Commission on Large Dams, ICOLD' played important roles in the worldwide spread and standardization of dam technology. ICOLD was founded in 1928 on occasion of an 'International Congress of Electricity Producers and Distributors' in Paris by delegates from France, Great Britain, Italy, Romania, Switzerland and the USA and it presently includes some 80 national committees. The 17th International Congress on Large Dams was held in 1991 in Vienna/Austria, and up to that time, over 3000 congress reports and many other publications were issued.

Worldwide standardization

International Commission on Large Dams

This exchange of knowledge also was instrumental for a considerable improvement of the safety of dams (Fig. 200). The poor performance of American dams (mostly embankments) at the beginning of the 20th century has already been mentioned and explained. Today, the risks are the same with embankment as with concrete dams (Schnitter 1976). Of some 100 breaks of structures of more than 15 m of height recorded worldwide and built since 1900, about half were caused by overtopping at times of flood. Roughly one fourth of the failures were due to defects either in the dam's foundation or in their bodies.

Improvement of dam safety

Figure 199. Dams over 15 m high built worldwide from 1950 to 1989.

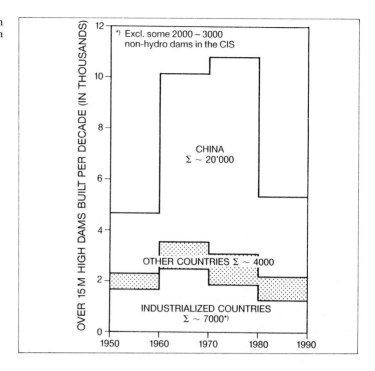

Figure 200. Failure of over 15 m high dams built since 1900 in the USA and Western Europe (eastern limit: Italy, Austria, Germany and Finland; without failures during construction, acts of war and tailings dams).

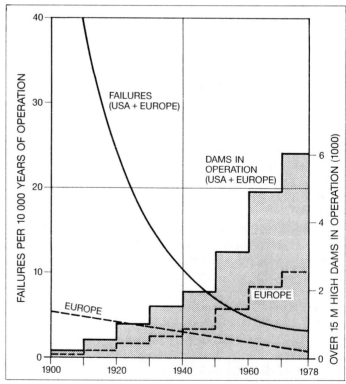

While the probability of failure diminished, the hazards created by dams increased because they grew in both number and size. So far, however, of the approximately 400 structures operative throughout the world and exceeding 100 m of height, none experienced irreparable trouble. Understandably, they were all designed, built and monitored with special care. Starting from the modest beginnings on embankment and arch dams mentioned before, the monitoring even of smaller dams became an art of its own. While it was relatively easy to measure uplift and pore pressures, it was difficult to directly check stresses and forces acting in the structures. Therefore, emphasis was given to the observation of their deformation by means of geodetic surveys, plumblines, settlement gauges etc. (Fig. 201). Additionally, some countries installed dam break alarms downstream of their most hazardous structures (Vischer 1982).

Along with the growth in size of the dams and their reservoirs their environmental impact and the awareness thereof increased as well (Table 49) (ICOLD 1973, 1984). In the late 1960's these aspects became the object of public debates, which, at times degenerated into bitter controversy (Williams & Veltrop 1991). In 1972, ICOLD set up a committee, which published a methodology for the environmental impact assessment of dam projects in 1980 (ICOLD 1980). Besides the change of landscapes – by no means always negative –, the disruption of fish migration and navigation as well as the not easily determinable effects on riverine plants and wildlife, dam constructions had drastic consequences, whenever large numbers of people had to be resettled from the reservoir area: up to 383 000 in the case of the Danjiangkou project, completed in 1974 some 900 km west of Shanghai (Cernea &

Importance of dam monitoring

Methods of monitoring

Environmental impacts

Resettlements

Figure 201. Geodetic observation of the deformations of the 145 m high Limmern arch dam completed in 1963 in eastern Switzerland (photo NOK, Baden/AG).

Table 49. Modern world record reservoirs.

Year of completion	Name of dam	Country (state)	Distance in km from city	Dam type	Height (m)	Reservoir[2] Capacity ($10^9 m^3$)	Area (km^2)
1902	Aswan	Egypt	700 S Cairo	Gravity	39	*1*	*120*
1912	Aswan[1]	Egypt	700 S Cairo	Gravity	44	*2.5*	*250*
1916	Boquilla	Mexico	140 SE Chihuahua	Gravity	74	*4.0*	210
1917	Gouin	Canada	320 N Montreal	Gravity	26	*8.6*	?
1936	Hoover (Boulder)	USA (Ariz./Nev.)	42 SE Las Vegas	Gravity	221	*34.9*	*660*
1937	Fort Peck	USA (Mont.)	450 NE Helena	Embankment	76	22.1	*970*
1941	Rybinsk	Russia	275 N Moscow	Gravity	30	25.4	*4550*
1955	Samara	Russia	800 SE Moscow	Gravity	45	*58.0*	*6150*
1959	Kariba	Zambia/Zimbabwe	130 S Lusaka	Arch	128	*160.4*	5100
1964	Bratsk	Russia	1170 E Novosibirsk	Buttress	125	*169.3*	5470
1965	Akosombo	Ghana	90 NE Accra	Embankment	134	148.0	*8480*

[1]After first heightening. [2]For comparison: Lake Geneva $90^9 m^3$ and 580 km^2.
Italics: Record breaking dimensions.

Moigne 1989). Unfortunately, several of these operations were poorly planned and/or badly managed. As is the case in other fields of technology it actually is not it that should be blamed, if anything goes wrong. To blame is the abuse of it that occurs for often shortlived and/or particular advantages.

7.2 OUTLOOK

Slow-down of dam construction...

During the last two decades, the number of newly built dams decreased considerably, lately also in China (Fig. 199). There, as in most industrialized countries, it was the consequence of a certain saturation and the running-out of economically viable dam sites, as e.g. in Switzerland (Fig. 202). In the other countries, dam construction was hindered by financial problems, although the need

...but increasing water needs

for more water supply, irrigation and flood control is obvious in the light of the rapidly growing world population, particularly where health and living standards ought to be increased at the same time. As shown in the left half of Figure 203, only some 9% of the annual runoff in the world's rivers of about 40 Tm^3 or almost 8000 m^3 per inhabitant and year are presently being diverted; 60% thereof being

Still ample opportunities

consumed, mostly by evapotranspiration in irrigation (Haws 1991). The rest is returned, alas, most often polluted. So there still seems to be plenty of fresh water around!

The trouble is that much of this flow occurs at the wrong time and therefore has to be regulated by means of reservoirs. This is

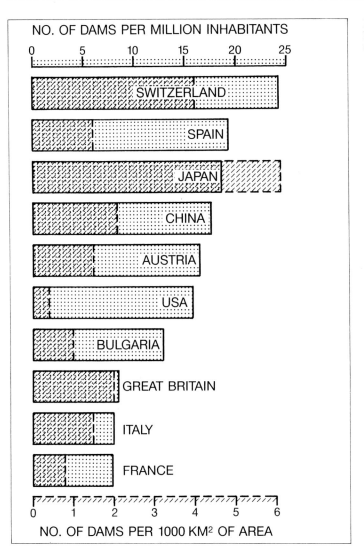

NO. OF DAMS PER MILLION INHABITANTS

Figure 202. Density of dams built up to 1989 in ten countries.

especially evident in the case of hydro-power, where in cold regions electricity consumption is highest in winter, when precipitation occurs mostly in the form of snow, while in hot regions energy for irrigation pumping and air-conditioning are required mainly during the dry season. As shown in the right half of Figure 203, the annual hydro-power potential of the world of 14 000 TWh or about 2700 kWh per person and year, today is also exploited to only 15% (WPDCH 1992). Even in the industrialized countries this percentage attains just 27% on average, although France, Bulgaria, Switzerland, the USA and Japan have already reached 97, 83, 75, 74 and 67% respectively. If the demand for electricity continues to grow – and there is no saturation in sight in this respect

Regulation by dams often mandatory

Large hydro-power potential

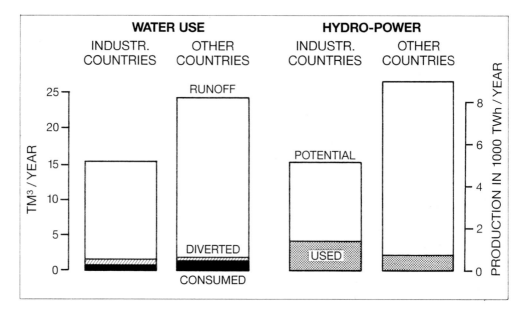

Figure 203. Present use of run-off and hydropower potential in the world.

Hydro-power is clean, renewable energy

Dam engineers ready to serve

– these countries, in order to adapt their increasing thermal (fossil and nuclear fuels) power production to consumption will have to build more pump-storage schemes.

They can also finance hydro-power development for exportation in Russia and Canada, where the per capita annual potential of 10 000 and 24 000 kWh respectively is well above the needs of the local population. In Canada, this is already done on a large scale, since almost half of its resources are developed, whereas in Russia, this rate stands at only 7%. Such a policy would certainly concur with the emphasis placed on the development of clean, renewable energy sources by the 1992 'United Nations Conference on Environment and Development, UNCED' in Rio de Janeiro in its Agenda 21, chapter 9. Whether the dams required to fulfill the demand ought to be large – they have the economies of scale and the lower number needed in their favour – or 'beautifully small', ultimately is a choice of the societies involved. Through a long history, dam engineers have acquired the ability to serve both goals efficiently, securely and in an environmentally safe way!

Chronology

Important dams and related inventions or publications in the order of their date of completion (with reference to the relevant pages in the book).

3000 BC	World's oldest dams in northern Jordan	(18-21)
2600 BC	Kafara dam in Egypt; failed during construction	(1-3)
2500 BC	First check dams in Baluchistan/Iran and Pakistan	(31)
1800 BC	Large reservoir in Faiyûm/Egypt depression	(4-5)
13th cent. BC	Mycenaean diversion and storage dams in Greece	(8-11)
950 BC	Solomon's Tyropoeon pool in Jerusalem	(21-22)
730 BC	Achaz's Probatica dam in Jerusalem	(22-23)
703 BC	First of Sennacherib's dams near Ninive/Iraq	(17-18)
700 BC	Urartian reservoirs in eastern Turkey	(15-16)
700 BC	Start of Purron dam in central Mexico	(46-47)
581 BC	Huge Anfengtang reservoir in central China	(41)
510 BC	Start of Sabean dam at Marib in Yemen	(25-27)
370 BC	First of many large reservoirs on Sri Lanka	(33)
275 BC	Largest Meroitic reservoir in northern Sudan	(7)
219 BC	Tianping divider weir in southern China	(44-45)
2nd cent. BC	Nabatean diversion dam at Petra in Jordan	(29)
1st cent. BC	So called 'Solomon pools' near Jerusalem	(23)
60 AD	First and highest (40 m) Roman dam near Rome	(58-59)
140	65 km long Jian (or Jin) embankment in central China	(41-42)

2nd cent.	Large Roman embankments in south-western Spain	(59-64)
270	Roman bridge-weirs in southwestern Iran	(76-78)
284	Large Roman gravity dam at Homs in Syria	(76)
380	Sayama embankment near Osaka in Japan	(106)
460	Paskanda Ulpotha embankment on Sri Lanka raised to 34 m	(36)
700	Palace reservoir at Tikal in Guatemala	(48-49)
7th/8th cent.	Several dams near Mekka and Medina in Saudi Arabia	(81-82)
970	Moslem Parada weir near Murcia in Spain	(84-85)
10th cent.	Bujid power dams in southern Iran	(86-87)
1030	Mahmud gravity dam in Afghanistan	(88-89)
1037	16 km long Veeranam embankment in southern India	(98)
1170	Parakrama's 'sea' on Sri Lanka	(95-96)
12th cent.	First mill dams and fish ponds in Europe	(107-109)
1272	Earliest of numerous fish ponds in Bohemia	(120-121)
1350	60 m high Kurit arch dam in central Iran	(91-93)
1384	Almansa curved gravity dam in south-eastern Spain	(124-126)
1404	First power dam in East German mining districts	(108)
1430	Dikes in Lake Texcoco around Mexico-City	(50-51)
1492	Jordán water supply embankment in Bohemia	(131)
1500	Castellar buttress dam in southwestern Spain	(108-109)
1547	Five books of Dubravius (J. Skála) on fish ponds	(123)
1558	First embankment with impervious core in German Ore Mountains	(109)
1560	First masonry water supply dam for Istanbul	(131-132)
1573	Earliest of several mining dams at Potosí in Bolivia	(145-146)
1578	13 000 million m^3 Hongze reservoir in central China	(104-106)
1594	46 m high Tibi curved gravity dam in southeastern Spain	(126)

1611	Start of Pontalto arch dam in South Tirol (completed 1887)	(144-145)
1632	First mill dam on Charles river in Massachussets	(147, 149)
1640	First modern arch dam of Elche in southeastern Spain	(126-127)
1675	36 m high St. Ferréol navigation embankment in France	(137-138)
1678	First embankments for Versailles park	(135)
1695	Joux Verte arched fluming dam in Switzerland	(142-143)
1723	Earliest mining dam in western Siberia	(115-116)
1735	Bedia multiple-arch dam in north-western Spain	(119-120)
1741	Ornamental lake embankment with clay core in Great Britain	(136)
1750	Several buttress dams in central Mexico	(147-148)
1765	Triangular gravity dams in central Mexico	(147-148)
1769	First of several fluming dams in Slovenia	(142-143)
1787	Start on 87 m high Gasco navigation dam near Madrid; construction stopped 1799 at 56 m; never in use	(138-139)
1797	First navigation embankments in Great Britain	(140-141)
1804	Mir Alam multiple-arch dam in central India	(151)
1831	Jones Falls arch dam in Ontario/Canada	(149-150)
1840	Entwistle embankment in Great Britain 38 m high	(158)
1852	Rockfill in South Fork dam in Pennsylvania	(159-161)
1853	J. de Sazilly's paper on triangular gravity dams	(170)
1854	Zola arch dam in France analysed with cylinder formula	(194-195)
1856	Parramatta thin arch dam in Australia	(195-196)
1866	60 m high Gouffre d'Enfer triangular gravity dam in France	(170-171)
1869	Hydraulic fill in Temescal dam in California	(161)
1872	Concrete in Boyds Corner/N.Y. gravity dam first time since Romans	(174)
1881	First concrete arch dam at Abbeystead in Great Britain	(197)

1882	42 m high Upper Barden embankment in Great Britain	(158)
1884	Big Bear Valley arch dam in California	(197)
1886	49 m high Lower Otay pure rockfill dam in California	(160, 162)
1889	Crown-cantilever analysis of arch dams	(198-199)
1903	Theresa/N.Y. reinforced-concrete, flat-slab buttress dam	(187-188)
1908	Hume Lake/Calif. reinforced-concrete, multiple-arch dam	(184-185)
1914	Salmon Creek variable-radius arch dam in Alaska	(199-200)
1915	105 m high Arrowrock concrete gravity dam in Idaho	(174)
1925	K. von Terzaghi's 'Principles of Soil Mechanics'	(164-165)
1928	Carranza contiguous buttress dam in Mexico	(191)
1929	119 m high Diablo arch dam in Washington	(201)
1931	100 m high Salt Springs rockfill dam in California	(169)
1936	221 m high Hoover gravity dam in Arizona/Nevada	(177)
1948	83 m high Escaba flat-slab buttress dam in Argentina	(190)
1957	237 (now 250) m high Mauvoisin arch dam in Switzerland	(207)
1961	285 m high Grande Dixence gravity dam in Switzerland	(178)
1968	214 m high D. Johnson multiple-arch dam in Canada	(187)
1968	230 m high Oroville embankment dam in California	(169)
1980	272 m high Inguri arch dam in Georgia	(201)
1980	300 m high Nurek embankment dam in Tajikistan	(169)
1981	Shimajigawa roller-compacted-concrete dam in Japan	(180)
1983	196 m high Itaipu contiguous buttress dam on Brazilian-Paraguayan border	(192-193)

References

Abrams, D.A.: *Design of Conrete Mixtures*. Structural Materials Research Lab., Chicago, 1918.

Adams, N.: Architecture for Fish, the Sienese Dam on the Bruna River, Structures and Design 1468 – ca. 1530. *Technology and Culture,* 1984, pp. 768-797.

Adams, R.M.: *Land Behind Baghdad*. Univ. Chicago Press, Chicago, 1965.

Agakhanian, G.A.: Armenia. *History of Irrigation and Drainage in the USSR* (B.G. Shtepa, ed.). Int. Com. Irrigation and Drainage, New Delhi, 1985, pp. 122-131.

Aird, W.V.: *The Water Supply, Sewerage and Drainage of Sydney*. Metrop. Water, Sewerage and Drainage Board, Sydney, 1961.

Almeida, D. Fernando de: Sobre a barragem romana de Olisipo e seu aqueducto. *O arqueólogo portugues*, 1969, pp. 179-189.

Ambursen, N.F. & Sayles, E.W.: A Hollow Concrete-Steel Dam at Theresa NY. *Eng. News*, 1903, part II, pp. 403-404.

Amiran, R.: The Water Supply of Israelite Jerusalem. *Jerusalem Revealed*. (Y. Yadin ed.) Yale Univ. Press, New Haven, 1967, pp. 75-78.

ANIDEL (Ass. naz. imprese produttrici e distributrici di energia elettrica): *Le dighe di ritenuta degli impianti idroelettrici italiani,* vol. 1. Milano, 1961.

Arenilla, M., Martin, J. & Alcaraz, A.: Nuevos datos sobre la presa de Proserpina. *Rev. obras publicas*, junio, 1992, pp. 65-69.

Armillas, P., Palerm, A & Wolf, E.R.: A Small Irrigation System in the Valley of Teotihuacan. *American Antiquity*, 1956, pp. 396-399.

ASCE: Report on Arch Dam Investigation. *Proc. Amer. Soc. Civil Engrs.*, May 1928, part 3.

Avistur, S.: On the History of the Exploitation of Water Power in Eretz-Israel. *Israel Exploration Journal*, 1960, pp. 37-45.

Bachmann, W.: Die assyrische Talsperre bei Ain-Siffni. *Wiss. Veröff. deutsche Orientges*. 1927, 52: 32-33.

Balcer, J.M.: The Mycenaean Dam at Tiryns. *Amer. Jour. of Archaeology*, 1974, pp. 141-149.

Balland, D.: Passé et présent d'une politique des barrages dans la région de Ghazni. *Studia Iranica*, 1976, pp. 239-253.

Bazant, Z.: Entwicklung der Berechnung von Bogenstaumauern. *2. Kongr. Int. Ver. Brücken-und Hochbau,* Berlin 1936, Vorbericht pp. 1107-1140.

Beier, H.: Talsperren in Bayern. *Geschicht. Entwicklung der Wasserwirtschaft und des Wasserbaus in Bayern*. Bayer. Landesamt für Wasserwirtschaft, München 1983, Teil 1, pp. 131-160.

Benoit, F.: Le barrage et l'aquéduc romains de Saint-Rémy de Provence. *Rev. études anciennes*, 1935, pp. 332-340.

Benson, R.P., Harland, W.P. & Pinkerton, I.L.: The Ancient Maduru Oya Sluiceway. *Int. Water Power & Dam Const.*, Dec. 1983, pp. 26-31.

Berdal, A.B.: Slab and Buttress Type Dams. *Concrete Dams in Norway.* Norconsult AS, Oslo 1968, pp. 5-8.

Bible (Old Testament): Ecclesiastes 2/6, 1st King 10 and 2nd Chronicles 9.

Binnie, G.M.: *Early Victorian Water Engineers.* T. Telford, London 1981.

Binnie, G.M.: *Early Dam Builders in Britain.* T. Telford, London 1987.

Biswas, A.K.: *History of Hydrology.* North-Holland Publ. Co., Amsterdam 1970.

Bittel, K. & Naumann, R.: *Bogazlöy-Hattusa.* W. Kohlhammer, Stuttgart, 1952.

Blatter, C.E. & Lendi, P.: *Der Injektionsschleier in Mattamrk.* Leemann, Zürich 1968.

Bligh, W.G.: The Ancient Irrigation and Watersupply Tanks or Reservoirs of Ceylon. *Eng. News*, Sept. 22, 1910, pp. 297-302.

Bohn, M. & Hamon, M.: The Djatiluhur Project. *Water Power*, 1967, pp. 305-315.

Bonnin, J.: Les hydrauliciens étrusques, des précurseurs? *La houille blanche*, 1973, pp. 641-649.

Botzan, M., Klein, R. & Vogel, A.: Historische Sperrenbauten in Rumänien. *Historische Talsperren,* Band 2. K. Wittwer, Stuttgart 1991, pp. 371-386.

Breznik, M.: The Safety and Endurance of the Old Dams of Idrija. *Safety of Dams.* A.A. Balkema, Rotterdam, 1984, pp. 133-139.

Brohier, R.L.: *Ancient Irrigation Works in Ceylon.* Ceylon Govt. Press, Colombo 1934-1935 (reprint: Min. of Mahaweli Devt., Colombo 1980).

Brossé, L.: La digue du lac de Homs. *Syria*, 1923, pp. 234-240.

Brown, J.M.: Contributions of the British to Irrigation Engineering in Upper India in the Nineteenth Century. *Trans. Newcomen Soc.*, 1983-1984, pp. 85-112.

Brunner, U.: Die Erforschung der antiken Oase von Marib mit Hilfe geomorphologischer Untersuchungsmethoden. *Archäol. Berichte aus dem Yemen.* Band II. Zabern, Mainz a.R. 1983.

Buffet, B. & Evrard, R.: *L'eau potable à travers les âges.* Solédi, Liège, 1950.

Bulota, G., Im, O. & Larivière, R.: Le barrage Daniel-Johnson, un vieillissement prématuré. *Congrès int. grands barrages,* Vienne 1991, vol. II, pp. 187-210.

Butzer, K.W.: *Early Hydraulic Civilization in Egypt.* Univ. Chicago Press, Chicago 1976.

Caballero Zoreda, L. & Sánchez-Palencia Ramos, F.J.: Presas romanas y datos sobre pobliamento romano y medieval en la provincia de Toledo. *Notic. Arqueol. hisp.*, no.14, 1982.

Caromb, Mairie de: *Lac du Paty.* Plaquette 1989 et plans de la Dir. Dép. de l'agriculture et de la forêt.

Castro Gil, J. de: El pantano de Proserpina. *Rev. obras publicas*, 1933, pp. 449-454 y 1936, pp. 203-204.

CDC: *Dams in Czechoslovakia.* Czechoslovak Dam Committee, Prague, 1967.

Çeçen, K.: Seldschukische und osmanische Talsperren. *Historische Talsperren,* Band 1. K. Wittwer, Stuttgart, 1987, pp. 275-295.

Celestino y Gomez, R.: Origines conceptuales de los complejos hidraulicos romanos en España; la presa romana de la Alcantarilla en Toledo. *Toletum*, no. 7, 1974, p. 161 ff. (drawing no. 3 is erroneous).

Celestino y Gomez, R.: Los sistemas romanos de abastecimiento de agua a Mérida. *Rev. obras publicas*, 1980, pp. 959-967.

Cernea, M.M. & Moigne, G.Le: The World Bank's Approach to Involuntary Resettlement. *Water Power & Dam Construction Handbook*. Reed Business Publ., Sutton/Surrey 1989, pp. 34-37.

CHINCOLD (Chinese Com. Large Dams): *Large Dams in China*. China Water Resources and Electric Power Press, Beijing 1987.

Coyne, A.: Arch Dams, their Philosophy. *Symp. Arch Dams*. Amer. Soc. Civil Engrs., New York, 1957, pp. 959/1-32.

Creager, W.P., Justin, J.D. & Hinds, J.: *Engineering for Dams*. J. Wiley & Sons, New York, 1945.

D'Andrea, D.: Gli impianti idroelettrici Volturno-Garigliano. *L'Elettrotecnica*, 1952, n. 1

Danilevskii, V.V.: *History of Hydroengineering in Russia before the 19th cent*. Israel Program Scient. Transl., Jerusalem 1968.

Davidson, I.: George Deacon (1843-1909) and the Vyrnwy Works. *Trans. Newcomen Soc.*, 1987-1988, pp. 81-95.

Davis, R.E.: Historical Account of Mass Concrete. *Symposium on Mass Concrete*. Amer. Concrete Inst., Detroit 1963, pp. 1-35.

Delocre, E.: Sur la forme du profil a adopter pour les grands barrages en maçonnerie des réservoirs. *Annales ponts et chaussées*, 1866, 2^{c}sem., pp. 212-272.

Denning, J.: Seismic Retrofitting. *Civil Eng.* (New York), Feb. 1993, pp. 48-51.

Díaz-Marta, M.: La ingeniería colonial en el Nuevo Mundo. *Rev. obras publicas*, 1981, pp. 695-699.

Díaz-Marta, M.: *Quatro obras hidraulicas antiguas*. Caja de ahorro de Toledo, Toledo 1992.

Doddiah, D.: Design of Siphon Spillways for Dams. *Int. Cong. Large Dams*, Paris 1955, Commun. 8.

Downey, G.: *Ancient Antioch*. Princeton Univ. Press, Princeton 1963, pp. 254-255.

Dunstan, M.R.H.: A Review of Roller Compacted Concrete Dams in the 1980's. *Int. Water Power & Dam Construction*, May 1990, pp. 43-45.

Eastwood, J.S.: An Arch Dam Design for the Shoshone Dam. *Eng. News*, 1910, part I, pp. 678-680.

Eastwood, J.S.: The New Big Bear Valley Dam. *Western Eng.*, 1913, pp. 458-470.

Ehrensperger, J.: Elektrizitätswerk am Löntsch. *Schweiz. Bauzeitung*, 1910 (Bd. 55 und 56).

ENEL (Ente Naz. Energia Elettrica): *Le dighe di ritenuta degli impianti idroelettrici italiani*. Roma 1970.

Engelbertsson, A.B.: Längs Salas vattenvägar. *Bergslags arkiv*, 1991, årg 3, pp. 3-17.

ENG.REC.: An Unusual Arched Dam. *The Eng. Record*, 1903, No. 2.

ENR: Two Arch Dams Fail Through Undermining of Abutments. *Eng. News Record*, Oct. 1926, pp. 616-618.

Erbiste, P.C.: Hydraulic Gates, the State-of-the-Art. *Int. Water Power & Dam Construction*, April 1981, pp. 43-48.

Evans, O.: *The Young Mill-Wright and Miller's Guide*. 1795 (Reprint: Oliver Evans Press, Wallingford/PA 1990).

Fahlbusch, H.: Alte Talsperren im Gebiet des Königreichs Saudi Arabien. *Historische Talsperren,* Band 1. K. Wittwer, Stuttgart, 1987, pp. 199-220.

Fecht, H.: Anlage von Stauweihern in den Vogesen. *Zeitschrift Bauwesen,* 1889.

Fellenius, W.: *Erdstatische Berechnungen.* W. Ernst & Sohn, Berlin/D 1926-1939.

Féret, L.R.: *Sur la compacité des mortiers hydrauliqes.* 1892.

Ferguson, P.A.S., Masurier, M. le & Stead, A.: Combs Reservoir; Measures taken following an Embankment Slip. *Int. Congr. Large Dams,* New Delhi 1979, vol. II, pp. 233-245.

Fernández Casado, C.: Las presas romanas en España. *Rev. obras publicas,* 1961, pp. 357-363.

Fernández Ordóñez, J.A., Abad Balboa, T., Andrés Mateo, C., Gálan Hergueta, A., A. & Martinez Vázquez de Parga, R.: *Catálogo de noventa presas y azudes españoles anteriores a 1900.* Com. Estud. Hist. Obras Publ. y Urbanismo (Cehopu), Madrid, 1984.

Fernando, A.D.N.: *The Ancient Hydraulic Civilization of Sri Lanka in Relation to its Natural Resources.* Royal Asiatic Soc., Sri Lanka Branch, Colombo, 1982.

Fleishman, T.: *Charles River Dams.* Charles River Watershed Assoc., Auburndale/MA, 1978.

Forchheimer, P.: *Hydraulik.* B.G. Teubner, Leipzig, 1914.

Forchheimer, P. & Strzygowski, J.: *Die byzantinischen Wasserbehälter von Konstantinopel.* Gerold & Comp., Wien 1893.

Frazier, A.H.: *Water Current Meters.* Smithsonian Inst. Press, Washington, 1974.

Fukuda, H.: Geschichte des Talsperrenbaus in Japan. *Historische Talsperren,* Band 2. K. Wittwer, Stuttgart 1991, pp. 315-327.

Fuller, W.E.: Flood Flows. *Trans. Amer. Soc. Civil Engrs.,* 1914, pp. 564.

Garbe, J.P., Biche, A., Gouvenot, D. & Guerrero, M.: Yacyreta, the Longest Dam in the World. *Travaux,* Mai 1991, pp. 80-86.

Garbrecht, G.: The Water Supply System at Tuşpa (Urartu). *World Archaeology,* 1980, pp. 306-312.

Garbrecht, G.: Sadd-el-Kafara, the World's Oldest Large Dam. *Int. Water Power and Dam Construction,* July 1985, pp. 71-76 (detaillierter Bericht: *Mitt. Leichtweiss-Inst. TU Braunschweig,* Heft 81, 1983).

Garbrecht, G.: Die Talsperren der Urartäer. *Historische Talsperren,* Band 1. K. Wittwer, Stuttgart 1987, pp. 139-145.

Garbrecht, G.: Neuere Ergebnisse von Untersuchungen über altägypptische Wasserbauten. *Frontinus-Hefte,* Nr 14, 1990, pp. 57-80 (detaillierter Bericht: *Mitt. Leichtweiss-Inst. TU Braunschweig,* Heft 107, 1990).

Garbrecht, G.: Talsperre und Tunnel am Hafen Seleukeia. *Historische Talsperren,* Band 2. K. Wittwer, Stuttgart 1991a, pp. 83-89.

Garbrecht, G.: Der Staudamm von Resafa-Sergiupolis. *Historische Talsperren,* Band 2. K Wittwer, Stuttgart 1991b, pp. 237-248.

Garbrecht, G. & Netzer, E.: Die Wasserversorgung des geschichtlichen Jericho und seiner Winterpaläste. *Bautechnik,* 1991, pp. 183-193.

Garbrecht, G. & Peleg, J.: Die Wasserversorgung geschichtlicher Wüstenfestungen am Jordantal. *Antike Welt,* 1989, Nr. 2, pp. 2-20.

Garbrecht, G. & Vogel, A.: Die Staumauern von Dara. *Historische Talsperren,* Band 2. K. Wittwer, Stuttgart 1991, pp. 263-276.

García-Diego, J.A.: Don Pedro Bernardo Villareal de Berriz y sus presas de contrafuertes. *Rev. obras publicas,* 1971 y 1972, pp. 233-238.

Gracía-Diego, J.A.: The Chapter on Weirs in the Codex of Juanelo Turriano. *Technology and Culture*, 1976, pp. 217-234.

García-Diego, J.A.: Old Dams in Extremadura. *History of Technology*, 1977, pp. 95-124.

García-Diego, J.A.: Nuevo estudio sobre la presa romana de Consuegra. *Rev. Obras publicas*, 1980, pp. 500-505.

García Tapia, N.: *Patentes de invencion españolas en el siglo de oro*. Min. industria y energia, Madrid 1990.

García Tapia, N. & Rivera Blanco, J.: La presa de Ontigola y Felipe II. *Rev. obras publicas*, 1985, pp. 479-492.

Garrandes, E.: La presa de El Gasco sobre el rio Guadarrama. *Bol. informacion minist. obras publicas*, Nov. 1973, pp. 20-25 y Dic. 1973, pp. 26-29.

Gauckler, P.: *Enquête sur les installations hydrauliques romaines en Tunisie*. Dir. antiquités, Tunis 1897-1902.

Gicot, H.: Conceptions et techniques de quelques barrages-voûtes suisses. *Wasser- und Energiewirtchaft*, 1961, pp. 194-205.

Glick, T.F.: *Irrigation and Society in Medieval Valencia*. Belknap, Cambridge/MA 1970.

Glossop, R.: The Invention and Development of Injection Processes. *Géotechnique*, 1960-1961, pp. 91-100 and 255-279.

Goblot, H.: Kébar en Iran, sans doute le plus ancien des barrages-voûtes. *Art et manufactures*, 1965, no. 154, pp. 43-49 (title of paper somewhat precipitious as acknowledged by author in: Sur quelques barrages anciens et la genèse des barrages-voûtes. *Rev. histoire des sciences*, 1967, pp. 109-140)

Goblot, H.: Du nouveau sur les barrages iraniens de l'époque mongole. *Art et manufactures*, 1973, no. 239, pp. 14-20.

Goldmark, H.: The Power Plant, Pipe-Line and Dam of the Pioneer Electric Power Company at Ogden/Utah. *Trans. Amer. Soc. Civil Engrs*, Dec. 1897, pp. 246-305.

Gomez-Navarro, J.L. & Aracil, J.J.: *Presas de Embalse*. Tipografía artística, Madrid 1958.

Graadt van Roggen, D.L.: Norice sur les anciens travaux hydrauliques en Susiane. *Mem. délégation scientif. en Perse*, 1905, tome 7, pp. 166-207.

Graeff, A.: Sur la forme et le mode de construction du barrage de Gouffre d'Enfer, sur le Furens, et des grands barrages en général. *Annales ponts et chaussées*, 1866, 2^{e}sem., pp.184-211.

Grewe, K.: Valle di Ariccia, Nemi-see, Albaner See (Italien). *Vermessungsing.*, 1981, pp. 203-206.

Grewe, K.: *Planung und Trassierung römischer Wasserleitungen*. Chmielorz, Wiesbaden 1985.

Grewe, K.: Wasserversorgung und -entsorgung im Mittelalter. *Die Wasserversorgung im Mittelalter*. P. von Zabern, Mainz a.R. 1991, pp. 9-86.

Gsell, S.: *Enquête administrative sur les travaux hydrauliques anciens en Algérie*. E. Leroux, Paris 1902.

Guenot & Thille: La stabilité des digues du réservoir et du contre-réservoir de Grosbois alimentant le canal de Bourgogne. *Annales ponts et chaussées*, 1949, pp. 467-491.

Hadfield, C.: *World Canals*. David & Charles, Newton Abbot, Devon 1986.

Hafner, F.: Bau und Verwendung von Triftklausen in Österreich vom 13. Jahrhundert bis zur Auflassung der Trift im 20. Jahrhundert. *Blätter für Technikgeschichte*, 1977-1978, pp. 47-64.

Hager, W.H.: Geschichte des Wassersprungs. *Schweiz. Ing. und Arch.*, 1990, pp. 728-735.

Harrison, C.L. & Woodard, S.H.: Lake Cheesman Dam and Reservoir. *Trans. Amer. Soc. Civil Engrs.*, 1904, pp. 89-209.

Hartung, F. & Kuros, G.R.: Historische Talsperren im Iran. *Historische Talsperren,* Band 1. K. Wittwer, Stuttgart 1987, pp. 221-274.

Häusler, E.: Zur Kolkproblematik bei Hochwasser – Entlastungsanlagen an Talsperren mit freiem Überfall. *Wasserwirtschaft*, 1980, Nr. 3.

Haws, E.T.: Environmental Issues in Dam Projects. *Int. Congr. Large Dams,* Vienna 1991, vol. I. pp. 1069-1191.

Hehmeyer, I.: Der Bewässerungslandbau auf der antiken Oase von Marib. *Archäol. Berichte aus dem Yemen.* Band V. P, von Zabern, Mainz a.R. 1991.

Helms, S.W.: *Jawa, Lost City of the Black Desert.* Cornell Univ. Press, Ithaca/ N.Y. 1981.

Herodotos of Halikarnassos: *History.* Book 2, chapters 5, 99, 101, 149 and 150 and Book 7, chapter 130.

Herzfeld, E.: *Geschichte der Stadt Samarra.* Eckardt & Messtorff, Hamburg 1948.

Hinds, J.: 200-Year-Old Masonry Dams in Use in Mexico. *Eng. News-Record*, 1932, pp. 251-253.

Hugentobler, W.: *Bericht der Kommission für Abdichtungen des Schweiz. Wasserwirtschaftsverbandes.* A. Bopp, Zürich 1927.

ICOLD: *World Register of Dams.* Int. Com. Large Dams, Paris 1973 and 1984.

ICOLD: *Dams and the Environment.* Int. Comm. Large Dams, Paris 1980.

Ilker, Ü. & Özbayoglu, Y.: A Major Grout Curtain in Semi-Karstic Limestone at Atatürk Dam. *Int. Congr. Large Dams,* Vienna 191, vol IV, pp. 687-696.

Imbertson, N.M.: Collapsible Dam aids Los Angeles Water Supply. *Civil Eng.* (New York), Sept. 1960, pp. 42-44.

INCOLD (Indian Nat. Com. Large Dams): *Major Dams in India;* and: *Register of Large Dams in India.* Both Central Board of Irrigation and Power, New Delhi 1979.

Jackson, D.C.: John S. Eastwood and the Mountain Dell Dam. *Jour. Soc. Industrial Archeology*, 1979, pp. 33-48.

Jacobsen, T. & Lloyd, S.: *Sennacherib's Aqueduct at Jerwan.* Univ. Chicago Press, Chicago 1935.

Jantzen, W.: *Führer durch Tiryns.* Deutsches archäol. Inst., Athen 1975.

JAPCOLD: *Dams in Japan 1984.* Japanese Com. Large Dams, Tokyo 1984.

Jorgensen, L.R.: The Constant-Angle Arch Dam. *Eng. News*, 1912, part II, pp. 155-157 and *Trans. Amer. Soc. Civil Engrs.*, 1915, pp. 685-733.

Judd, N.M.: The Material Culture of Pueblo Bonito. *Misc. Coll.* No. 124, Smithsonian Inst., Washington 1954.

Kalcyk, H.: *Untersuchungen zum attischen Silberbergbau.* P. Lang, Franfurt a.M. 1982.

Kedar, Y.: Water and Soil from the Desert, some Ancient Agricultural Achievements in the Central Negev. *Geographical Journal*, 1957, pp. 179-187.

Kincaid, W.: Rambles among Ruins in Central India. *Indian Antiquary*, 1888, pp. 348-352.

Kleinschroth, A.: Staudämme im altsudanesischen Reich Kush (Nubien). *Historische Talsperren,* Band 1. K. Wittwer, Stuttgart 1987, pp. 123-137.

Knauss, J.: Die Melioration des Kopaisbeckens durch die Minyer im 2. Jt.v.Chr. *Berichte Inst. Wasserbau......TU München*, Nr. 57, 1987.

Knauss, J.: Die alten Talsperren beim taubenumschwärmten Thisbe in Südwestböotein. *Antike Welt*, 1989, Nr. 3, pp. 35-55.

Knauss, J.: Mykenische Wasserbauten in Arkadien, Böotien und Thessalien; mutmassliche Zielsetzung und rekonstruierbare Wirkungsweise. *Frontinus-Hefte*, Nr. 14, 1990, pp. 17-56.

Kottmann, A.: *Alte Baumaschinen*. Schell & Steiner München 1979.

Kratochvil, S.: Historie výstavby prvníc prehrad v Čechách a na Moravê. *Vodní hospodářství*, 1967, pp. 261-265.

Lamprecht, H-O.: *Opus Caementitium*. Beton-Verlag, Düsseldorf, 1984.

Lanser, O.: Die Anfänge des österreichischen Talsperrennbaues. *Blätter für Technikgeschichte*, 1960, Nr. 22, pp. 150-171 und: *Statistik der österreichischen Talsperren*. Öster. Wasserwirtschaftsverband, Wien 1962, pp. 7-25.

Leach, E.R.: Hydraulic Society in Ceylon. *Past and Present*, April 1959, pp. 1-26.

Leffel, J.: *Construction of Mill Dams*. Springfield/OH 1881 (reprint: Noves, Park Ridge/NJ 1972).

Legget, R.F.: The Jones Falls Dam on the Rideau Canal, Ontario/Canada. *Trans. Newcomen Soc.*, 1957-1959, pp. 205-218 and *Rideau Waterway*. Univ. Toronto Press, Toronto/Canada 1955 and 1972.

Liere, W.J. van: Traditional Water Management in the Lower Mekong Basin. *World Archaeology*, 1980, pp. 265-280.

Lindner, M.: Nabatäische Talsperren. *Historische Talsperren,* Band 1. K. Wittwer, Stuttgart 1987, pp. 147-174.

Loriferne, H.: Aperçu historique sur le service des eaux et fontaines de Versailles, Marly et Saint-Cloud et son évolution. *Techniques et sciences municipales,* Jan. 1963, pp. 13-38.

Malhotra, O.P.: Epic Repair Battle at Bhakra Dam. *World Construction*, Aug. 1960, pp. 17-18, 34 and 37-39.

Malinowski, R. & Garfinkel, Y.: Prehistory of Concrete. *Concrete International*, March 1991, pp. 62-68.

Mansergh, J.: Lancaster Waterworks Extension. *Proc. Inst. Civil Engrs.* 1881-1882, part II, pp. 253-277.

Marcello, C.: Considérations sur les exemples réalisés d'un type de barrage à éléments évidés. *Congr. int. grands barrages,* Paris 1955, vol. IV, pp. 1479-1502.

Martin, J.B. & Muñoz Bravo, J.: *Las presas del estrecho de Puentes*. Conf. hidrograf. Segura, Murcia//E 1986.

Marx, C.: Entwicklung der Bohrtechnik. *Frontinus Hefte*, Nr. 14, 1990, pp. 125-150.

Mary, M.: *Barrages-voûtes; historique, accidents et incidents*. Dunod, Paris 1968.

Mason, R.D., Lewarch, D.E., O'Brien, M.J. & Neely, J.A.: An Archaeological Survey of the Xoxocotlan Piedmont, Oaxaca/Mexico. *American Antiquity*, 1977, pp. 567-575.

Matthies, A.L.: The Medieval Wheelbarrow. *Technology and Culture*, 1991, pp. 356-364.

Mazar, A.: Survey of the Jerusalem Aqueducts. *Mitt. Leichtweiss-Inst. TU Braunschweig*, Nr. 82, 1984, 27 pp.

McCullough, D.G.: *The Johnstown Flood*. Simon and Schuster, New York, 1968.

Mezquiriz, M.A.: Comentario al estudio conjunto sobre la presa romana de Consuegra. *Rev. obras publicas*, 1984, pp. 194-199.

Millon, R.: Irrigation Systems in the Valley of Teotihuacan. *American Antiquity*, 1957, pp. 160-166.

MOC (Ministry of Construction): *Construction of Shimajigawa Dam*. Hiroshima, undated.

Moore, E.: Water Management in Early Cambodia; Evidence from Aerial Photography. *Geographical Jour.*, 1989, pp. 204-214.

Mukhamedjanov, A.R.: Popular Irrigation Practices and Water Economy of Turkestan before it became Part of Russia. *History of Irrigation and Drainage in the USSR* (B.G. Shtepa ed.). Int. Com. Irrigation and Drainage, New Delhi 1985, pp. 80-93.

Murray, W.M.: The Ancient Dam of the Mytikas Valley. *Amer. Journal Archaeology*, 1984, p. 195-203.

Murti, N.G.K.: State of the Art of Earth Backing of Masonry Dams. *Irrigation and Power*, 1979, pp. 351-366.

Naumann, R.: *Architektur Kleinasiens, von ihren Anfängen bis zum Ende der hethitischen Zeit*. E. Wasmuth, Tübingen 1955 und 1971.

Needham J.: *Science and Civilisation in China*. Cambridge Univ. Press, London 1971, vol. 4, part III, pp. 211-378.

Neely, J.A.: Sassanian and Early Islamic Water-Control and Irrigation Systems on the Deh Luran Plain, Iran. *Irrigation's Impact on Society*. (T.E. Downing and M. Gibson eds.) Univ. Arizona Press, Tucson 1974, pp. 21-42.

Nimmo, W.H.R.: Historical Review of Dams in Australia. *Bull. Austral. Nat. Com. Large Dams*, Jan. 1966, pp. 11-51.

Noetzli, F.A.: Improved Type of Multiple-Arch Dam. *Trans. Amer. Soc. Civil Engrs.*, 1924, pp. 342-413.

Noetzli, F.A.: Don Martin Dam, Mexico. *Western Construction News*, 1928.

Noetzli, F.A.: Pontalto and Madruzza Arch Dams. *Western Constr. News and Highway Builder*, 1932, pp. 451-452.

Novak, J.: Vznik vodných nádrží v okoli Banskej Stiavnice a ich význam do začiatku 20. storočia. *Vodní hospodářství*, 1972, pp. 198-202.

O'Brien, M.J., Lewarch, D.E., Mason, R.D. & Neely, J.A.: Functional Analysis of Water Control Features at Monte Alban, Oaxaca/Mexico. *World Archaeology*, 1980, pp. 342-355.

Oleson, J.P.: Technology and Society in Ancient Edom, the Humayma Hydraulic Survey 1986. *Trans. Royal soc. Canada*, 1987, pp. 163-174.

Osten, H.H. von der, Martin, R.A. & Morrison, J.A.: *Discoveries in Anatolia 1930-1931*. Univ. Chicago Press, Chicago 1933.

Özis, Ü.: The Ancient Dams of Istanbul. *Int. Water Power & Dam Const.*, July 1977, pp. 49-51 and Aug. 1977, pp. 44-47.

Palerm, A.: *Obras hidraulicas prehispanicas en el sistema lacustre del valle de Mexico*. Inst. nac. antropologia e historia. Mexico 1973.

Parker, H.: *Ancient Ceylon*. Luzac & Co., London 1909.

Peleg, J.: Das Stauwerk für die untere Wasserleitung nach Caesarea. *Mitt. Leichtweiss-Inst. TU Braunschweig*, Nr. 89, 1986, 25 pp.

Peleg, J.: Geschichtliche Talsperren im Negev. *Historische Talsperren*, Band 2. K. Wittwer, Stuttgart 1991, pp. 101-108.

Pelletreau, A.: Barrages cintrés en forme de voûte. *Annales ponts et chaussées*, 1879, 1er sém., pp. 198-218.

Penman, A.D.M.: Materials and Construction Methods for Embankment Dams and Cofferdams. *Int. Congr. Large Dams,* Rio de Janeiro 1982, vol IV, pp. 1105-1228.

Peter, P. & Lukác, M.: Stiavnický systém vodných nádrží z hl'adiska dnešných požiadaviek na ich konštukciu a spôsob využívania. *Vodní hospodářství*, 1972, pp 203-207.

Petterson, K.E.: The Early History of Circular Sliding Surfaces. *Géotechnique*, 1955, pp 275-296.

Pierre, M-J. & Rousée, J-M.: Sainte-Marie de la Probatique; état et orientation des recherches. *Proche-Orient Chrétien*, 1981, pp. 23-42.

Plutarch: *Vitae parallelae,* Marcellus 17.

Poláček, J. & Dvořáčková, J.: Checking of Ageing (of Dams), Methods and Results. *Int. Cong. Large Dams,* Vienna 1991, vol II, pp. 531-539.

Procopius of Caesarea: *Buildings* (of Justinian). Book II, chapters 3, 7 and 10.

Quintela, A. de Carvalho, Cardoso, J.L. & Mascarenhas, J.M.: *Aproveitamentos hidráulicos romanos a sul do Tejo.* Dir. ger. recursos e aproveitamentos hidráulicos, Lisboa 1986 (English summary: *Int. Water Power & Dam Constr.,* May 1987, pp. 38-40 and 70).

Quintela, A. de Carvalho, Cardoso, J.L. & Mascarenhas, J.M.: *Barragens antigas em Portugal a sul do Tejo.* Fund. San Benito de Alcantara, Alcantara/E. 1988.

Raikes, R.L.: The Ancient Gabarbands of Baluchistan. *East and West,* 1964-1965, pp. 26-35.

Rao, K.L.: Earth Dams, Ancient and Modern, in Madras State. *Int. Cong. Large Dams,* New Delhi 1951, vol. I, pp. 285-301.

Rao, K.L.: Stability of Slopes in Earth Dams and Foundation Excavations. *Int. Conf. Soil Mech. and Found. Eng.,* Paris 1961, vol II, pp. 691-695.

Raphael, J.M.: Hoover Dam Plus Fifty Years Equals El Cajon Dam. *Waterpower '85.* Amer. Soc. Civil Engrs., New York 1985, pp. 1278-1287.

Rec. Hydr.: *Presas construidas en Mexico.* Secr. recursos hidraulicos, Mexico 1976.

Reynolds, T.S.: *Stronger than a Hundred Men.* J. Hopkins Univ. Press, Baltimore 1983.

Reynolds, T.S.: A Narrow Window of Opportunity, the Rise and Fall of the Fixed Steel Dam. *Jour. Soc. Industrial Archeology,* 1989, pp. 1-20.

Reza, E., Kuros, G., Emam, M. & Entezami, A.: *Water and Irrigation in Ancient Iran* (in Iranian). Iran Chap, Tehran 1971.

Rhone, T.J.: Problems concerning Use of Low Head Radial Gates. *Jour. Hydraulics Div. Amer. Soc. Civil Engrs.,* Feb. 1959, pp. 35-65.

Rigaud, J.: *L'ingénieur François Zola, sa famille, sa vie et son oeuvre.* Académie, Aix-en-Provence 1957.

Ritter, H.: *Die Berechnung von bogenförmigen Staumauern.* J. Lang, Karlsruhe 1913.

Robin, C., Breton, J-F. & Audouin, R.: Un patrimoine menacé. *Archeologia,* No. 160, 1981, pp. 36-43.

Rohn, A.H.: Prehistoric Soil and Water Conservation on Chapin Mesa, Southwestern Colorado. *American Antiquity,* 1963, pp. 441-455.

Rolt, L.T.C.: *From Sea to Sea.* A. Lane, London 1973.

Rothmund, H.: Entwicklung und Gestaltung von Saugüberfällen. *Der Bauingenieur,* 1962, pp. 135-138.

Rouse, H. & Ince, S.: *History of Hydraulics.* Iowa Inst. Hydr. Research, Iowa City 1957.

Rudolph, W.E.: The Lakes of Potosí. *Geographical Review,* 1936, pp. 529-554.

Saladin, H.: Description des antiquités de la régence de Tunis. *Arch. missions scient. et litt.,* serie 3, vol. 13, 1886 (pp. 162-163).

Sandström, G.A.: *The History of Tunnelling.* Barrie and Rockliff, London 1963.

Sazilly, J.A. Torterne de: Sur un type de profil d'égale résistance proposé pour les murs de réservoirs d'eau. *Annales ponts et chaussées,* 1853, pp 191-222.

Schlumberger, D.: Les fouilles de Qasr el-Heir el-Gharbi. *Syria*, 1939, pp. 195-238 et 324-373.

Schmidt, M.: Die Wasserwirtschaft des Oberharzer Bergbaues. *Frontinus-Hefte*, Nr. 13, 1989.

Schmidt, M. & Hobst, L.: Frühe Teichbautechnik in Böhmen. *Wasser und Boden*, 1991, pp. 602-605.

Schneider, A.: Die drei ehemaligen herzoglich-württembergischen Fischweiher bei Nabern. *Denkmalpflege in Baden-Württemberg*, 1989, pp. 192-197.

Schnitter, N.: The Emosson Arch Dam. *Water Power*, March 1974, pp. 77-87.

Schnitter, N.: Statistische Sicherheit der Talsperren. *Wasser, Energie, Luft*, 1976, pp. 126-129.

Schnitter, N.: Koloniale Aquädukte in Mexico. *Frontinus-Hefte*, Nr. 5, 1982, pp. 71-81.

Schnitter, N.: Altgriechischer Wasserbau. *Schweiz. Ing. u. Arch.*, 1984, pp. 479-486.

Schnitter, N.: El Cajon Arch Dam. *Concrete Int.*, Aug. 1984, pp. 7-13.

Schnitter, N.: Die Geschichte des Wasserbause in der Schweiz. Olynthus Verlagsanstalt, Vaduz/FL 1992.

Schrader, E.K.: The First Concrete Gravity Dam Designed and Built for Roller Compacted Construction Methods. *Concrete Int.*, Oct. 1982, pp. 15-24.

Schulze, O.: Notes on the Belubula Dam. *Trans. Australian Inst. Mining Engrs.*, 1897, pp. 161-172 (Summary: *Eng. News*, Sept. 8, 1898).

Schuyler, J.D.: The Construction of the Sweetwater Dam. *Trans. Amer. Soc. Civil Engrs.*, 1888, pp 201-232.

Segal, J.B.: *Edessa, the Blessed City.* Clarendon, Oxford 1970, pp. 187-188.

Semenza, C.: Arch Dams Development in Italy. *Symp. Arch Dams*. Amer. Soc. Civil Engrs., New York 1957, pp. 1017/1-42.

Sherman, L.K.: Streamflow from Rainfall by the Unitgraph Method. *Eng. News Record*, 1932 (vol. 108), pp. 501-505.

Shrava, S.S.: *Irrigation in India Through Ages.* Central Board of Irrigation & Power, New Delhi 1951.

Sieber, H.U.: *Talsperren in Sachsen.* Landestalsperrenverwaltung, Pirna/D 1992.

Siewert, H.H.: Bauten der Wasserwirtschaft im Yemen. *Baghdader Mitt. deutsches archäol. Inst.*, 1979, pp. 168-178.

Siewert, H.H.: Antike Bewässerungsbauten der yemenitischen Landwirtschaft. *Archäol. Berichte aus dem Yemen*, Band I. P. von Zabern Mainz a.R. 1982, pp. 181-196.

Skempton, A.W.: Alexandre Collin. *Trans. Newcomen Soc.*, 1946, pp. 91-104 and *Géotechnique*, 1949, pp. 215-222.

Skempton, A.W.: *Early Printed Reports and Maps (1665-1850) in the Library of the Institution of Civil Engineers.* Inst. Civil Engrs., London 1977.

Skempton, A.W.: Landmarks in Early Soil Mechanics. *7th Europ. Conf. Soil Mech. and Found. Eng.*, Brighton 1979, vol. 5, pp. 1-26. and *11th Int. Congr. Soil Mech. and Found. Eng.*, Jubilee vol., A.A. Balkema, Rotterdam 1985, pp. 91-118.

Smith, N.A.F.: The Roman Dams of Subiaco. *Technology and Culture*, 1970, pp. 58-68.

Smith, N.A.F.: *A History of Dams.* P. Davies, London 1971.

Smith, N.A.F.: Attitudes to Roman Engineering and the Question of the Inverted Siphon. *History of Technology*, 1976, pp. 45-71.

Stark, H.: Geologische und technische Beobachtungen an alten anatolischen Talsperren. *Wasserwirtschaft*, 1957-1958, pp. 16-19.

Stein, A.: Surveys on the Roman frontier in Iraq and Trans-Jordan. *Geogr. Journal*, 1940, pp. 428-438.

Stucky, A.: Etude sur les barrages arqués. *Bull. tech. suisse romande*, 1922, pp. 1-7, 25-30, 49-53, 85-90 et 97-103.

Stucky, A.: Le barrage de la Dixence. *Bull. tech. suisse romande,* 1946, pp. 37-48, 53-64 et 97-105.

Stucky, A.: Barrages-voûtes en Suisse. *Wasser- und Energiewirtschaft*, 1956, pp. 230-240.

Talobre, J.A.: *La mécanique des roches.* Dunod, Paris 1957.

Thompson, R.C. & Hutchinson, R.W.: The Agammu (Pool or Reservoir) of Sennacherib on the Khosr. *Archaeologia*, 1929, pp. 114-116.

Time Magazine, New York Oct. 25, 1963, pp 42-43.

Timoshenko, S.P.: *History of Strength of Materials.* McGraw-Hill, New York 1953.

Tolle, J.M., Simard, P.W. & Brown, L.A.: Modern Engineering Saves Troubled Dam. *Civil Eng.* (New York), June 1979, pp. 78-80.

Toulouse, J.H.: Early Water Systems at Gran Quivira National Monument. *American Antiquity*, 1945, pp. 362-372.

Tournadre, de: Note A annexée à la description d'un barrage exécuté sur la rivière Verdon....*Annales ponts et chaussées*, 1872, 1er sém., pp. 428-455.

USBR: *Model Studies of Spillways.* US. Bureau of Reclamation, Denver 1938a.

USBR: *Trial Load Method of Analyzing Arch Dams.* US Bureau of Reclamation, Denver 1938b.

USCOLD (US. Com. Large Dams): *Lessons from Dam Incidents,* USA, 2 vol. Amer.Soc. Civil Engrs., New York 1975 and 1988.

USCOLD (US. Com. Large Dams): *Development of Dam Engineering in the United States.* Pergamon, New York 1988.

Vadera, S.H.L.: *Development of Irrigation in India.* Central Board of Irrigation & Power, New Delhi 1965.

Vercoutter, J.: The Flooded Fortresses of Nubia. *UNESCO-Courier*, 1980, No. 2/3.

Villareal de Berriz, P.B.: *Maquniaz hhidraulicas de molinos y herrerias y govierno de los arboles y montes de Vizcaya.* A. Marin, Madrid 1736 (reprint: Soc. Guipuzcoana, San Sebastian 1973).

Vischer, D.L.: Water Alarm Organization in Zürich. *Int. Congr. Large Dams*, Rio de Janeiro 1982, vol I, pp 993-1000.

Vischer, D.: Die Beiträge bedeutender Schweizer Pioniere im Gebiet der Hydrologie und der Hydraulik. *Die Geschichte der Gewässerkorrektionen und der Wasserkraftnutzung in der Schweiz*, Pro Aqua, Basel 1983, pp. 0.3-0.26.

Vischer, D. & Schnitter, N.: *Drei Schweizer Wasserbauer.* Ver. wirtschaftshist. Studien, Meilen/ZH 1991.

Vischer, H. & Wagoner, L.: On the Strains in Curved Masonry Dams. *Trans. Tech. Soc. Pacific Coast*, Dec. 1889, pp. 75-151.

Vita-Finzi, C.: Roman Dams in Tripolitania. *Antiquity*, 1961, pp. 14-20.

Vita-Finzi, C. & Brogan, O.: Roman Dams on the Wadi Megenin. *Libya antiqua*, 1965, pp. 65-71.

Votruba, L.: Historische Talsperren und Wasserseicher in der CSSR. *Historische Talsperren,* Band 1. K. Wittwer, Stuttgart 1987, pp. 401-406.

Votruba, L.: Die Wasserversorgungstalsperre Jordan in Südböhmen vom Jahr

1492. *Historische Talsperren,* Band 2. K. Wittwer, Stuttgart 1991, pp. 341-352.

Vries, B. de: The el-Lejjun Water System. The Roman Frontier in Central Jordan (S.T. Parker ed.). *Brit. Archaeol. Rep., Int Series,* 1987, part 1, pp. 399-428.

Wade, L.A.B.: Concrete and Masonry Dam Construction in New South Wales. *Proc. Inst. Civil Engrs.,* 1908-1909, part IV, pp. 1-110.

Wagenbreth, O.: Historische Anlagen der bergmännischen Wasserwirtschaft im sächsischen Erzgebrige. *Frontinus-Hefte,* Nr. 15, 1991, pp. 113-133.

Wegmann, E.: *The Design and Construction of Dams.* J. Wiley & Sons, New York 1888-1927.

White, L.: *Medieval Religion and Technology.* Univ. California Press, Berkeley 1978.

Wilkinson, J.: The Pool of Siloam. *Levant,* 1978, pp. 116-125.

Williams, G.S.: Discussion on Lake Cheesman Dam and Reservoir. *Trans. Amer. Soc. Civil Engrs.,* 1904, pp. 182-185.

Williams, P.B. & Veltrop, J.: The Debate over Large Dams. *Civil Eng.* (New York), Aug. 1991, pp. 42-48.

Wisner, G.Y. & Wheeler, E.T.: Investigation of Stresses in High Masonry Dams of Short Spans. *Eng. News,* 1905, part II, pp. 141-144.

Wissmann, H. von: *Zur Landeskunde des Yemen.* Kümmerly & Frey, Bern 1966, pp. 93-108.

Woodbury, R.B. & Neely, J.A.: Water Control Systems of the Tehuacan Valley. *The Prehistory of the Tehuacan Valley.* (F. Johnson ed.) Univ. Texas Press, Austin 1972, vol. 4, pp. 81-153.

Woolley, C.L. & Lawrence, T.E.: The Wilderness of Zin. *Palestine Exploration Fund Ann.,* 1914-1915, pp. 121-128.

WPDCH: The World's Hydro Resources. *Water Power & Dam Construction Handbook.* Reed Business Publ., Sutton/Surrey 1992, pp. 34-43.

Zheng Liandi: Talsperren und Wehre im alten China. *Historische Talsperren,* Band 2. K. Wittwer, Stuttgart 1991, pp. 109-129.

Ziegler, P.: *Der Talsperrenbau.* A. Seydel, Berlin 1900 und W. Ernst, Berlin 1911 und 1925.

Index

Names of people and persons are in CAPITALS, subjects are in *italics*.